W9-BCF-280

ANNALS OF MATHEMATICS STUDIES
NUMBER 12

MEROMORPHIC FUNCTIONS AND ANALYTIC CURVES

BY

HERMANN WEYL

In collaboration with

F. JOACHIM WEYL

PRINCETON
PRINCETON UNIVERSITY PRESS
LONDON: HUMPHREY MILFORD
OXFORD UNIVERSITY PRESS
1943

Copyright 1943
PRINCETON UNIVERSITY PRESS

Lithoprinted in U.S.A.
EDWARDS BROTHERS, INC.
ANN ARBOR, MICHIGAN
1943

MATH.-SCI.
469156
QA
331
N54m

PREFACE

Five years ago my son Joachim and I discovered and brought home from the primeval forest of mathematics, a sapling which we called Meromorphic Curves (Annals of Mathematics, 1938). It looked healthy and attractive, but we did not know much about it. Soon after, a gardener from the North came along, - a skillful man of great experience, L. Ahlfors was his name, he knew; and under his care the plant, almost overnight, grew into a beautiful tree (Acta Soc. Sci. Fenn., 1941). Having learned the lesson, we set ourselves to carry out an idea dimly conceived before (Annals of Math. 1941, Proc. Nat. Acad. 1942), namely to transplant the tree Meromorph from the z-plane into the mountainous terrain of an arbitrary Riemann surface (a landscape of which I have been fond since the early days of my youth). The experiment seems to have succeeded. The leaves are out, a few buds are visible, but only the future can teach what fruits the tree will bear. In the meantime the howling storm of war has cut us off from our wise gardener.

The material of this "Study", including many details of design, shaped itself in conversations and correspondence with my son; in this sense it is a joint enterprise. But the bulk of the manuscript was actually written down by the undersigned as hour by hour notes of a course given in the Institute for Advanced Study during the first term of 1942-1943, and responsibility for the final arrangement and text rests with him.

I wish to acknowledge valuable assistance rendered by

G1731l - UBGS - 6/1/45 - 3.15
UNIVERSITY LIBRARY ROCHESTER, N.Y.

Dr. Audrey Wishard McMillan and Mr. Fumio Yagi in preparing the manuscript, and helpful criticisms and suggestions offered during the course by Professor Claude Chevalley.

Hermann Weyl

April 1943
Institute for Advanced Study
Princeton, New Jersey

CONTENTS

(4.12) refers to the formula labeled (4.12) in §4 of the same Chapter.

(II, 4.12) refers to formula (4.12) of Chapter II.

INTRODUCTION

EARLY HISTORY AND BIBLIOGRAPHY

The meromorphic functions bear the same relation to the entire functions as the rational functions do to the polynomials. The most important characteristic of a polynomial $f(z)$ is its <u>degree</u>. It has a two-fold significance: it counts the number of zeros and it describes the growth of $|f(z)|$ with increasing $|z|$. The only polynomials without zeros are the constants, and a polynomial with the n zeros $a_1, \ldots, a_n$ (each counted with its proper multiplicity is of the form

$$C \cdot \prod_i (z-a_i) \quad \text{or} \quad C^* \cdot \prod_i \left(1 - \frac{z}{a_i}\right).$$

The second form presupposes that all roots $a_i \neq 0$. If h of them equal zero, one must write instead

$$C^* z^h \cdot \prod_i \left(1 - \frac{z}{a_i}\right)$$

with the product now extending only over the n - h roots a_i which are different from zero. The type of growth of a polynomial of degree n is either described by an inequality

$$|f(z)| \leq \text{Const.}\ |z|^n$$

or, more precisely, by an asymptotic equation

$$(1) \qquad |f(z)|/|z|^n \to C \neq 0 \qquad \text{for } |z| \to \infty .$$

The double role of the degree n is perhaps better understood if we interpret (1) to mean that $f(z)$ takes on the value ∞ with multiplicity n at $z = \infty$. The resulting law, that a polynomial assumes the value 0 as often as ∞, carries over at once from polynomials to their quotients, the rational functions $f(z)$; and if one applies the law to $f(z) - c$ instead of to $f(z)$ it asserts that a rational function assumes each value c, including ∞, the same number of times. The validity of the law requires that one include the value $z = \infty$ as well as the finite values in the range of the argument ("z-sphere" instead of z-plane).

Can we make statements about entire functions which parallel our above observations about polynomials? Searching for analogies, we experience a serious shock right at the beginning: there are lots of entire functions without zeros. The classical example of the exponential function e^z is at the same time, in a certain sense the typical one. Indeed, every entire function $f(z)$ without zeros is of the form $e^{\phi(z)}$ where $\phi(z)$ is entire. We prove this either by applying to $\log f(z)$ the monodromy theorem according to which analytic continuation of a function element gives rise to a single-valued function of z, if the process encounters no obstacle in the z-plane; or by applying Cauchy's integral theorem to the logarithmic derivative $f'(z)/f(z)$ of f. The latter argument runs as follows: For a regular f without zeros, the integral

$$\int_0^z \frac{f'(z)}{f(z)}\, dz$$

defines a function $\phi(z)$ which is single-valued and regular everywhere. Since $f(z) \cdot e^{-\phi(z)}$ has the derivative zero, $f(z)$ equals Const. $e^{\phi(z)}$.

Given any entire function $f(z)$, each circle $|z| \leqq R$ contains but a finite number of its zeros. Hence we may arrange the zeros a_n, counting each with its proper multiplicity, by increasing moduli $|a_n|$, so that

$$|a_1| \leqq |a_2| \leqq |a_3| \leqq \cdots, \qquad |a_n| \to \infty \text{ with } n \to \infty$$

We ask conversely: Given a sequence a_n such that

$|a_n| \to \infty$ with $n \to \infty$, can we construct an entire function of which these a_n are the zeros? The answer is affirmative and the construction was carried out by Weierstrass (1876). His paper is the starting point of many subsequent investigations on entire and meromorphic functions. Of course it would now be foolish to form the product $\prod(z-a_n)$ but one might try our second form

(2) $$f_0(z) = \prod(1 - \frac{z}{a_n}).$$

Under the assumption that the series

(3) $$\sum |a_n|^{-1}$$

converges this "canonical product" is uniformly convergent in any given circle $|z| \leq R$ and hence yields an entire function with the preassigned zeros a_n. If the zeros are more densely spread, Weierstrass is inspired by the power series

$$- \log (1-u) = \frac{u}{1} + \frac{u^2}{2} + \frac{u^3}{3} + \ldots$$

to form the "primary factor"

$$E(k;u) = (1-u) \cdot \exp (\frac{u}{1} + \ldots + \frac{u^k}{k}).$$

He then finds that one may fix integers $k_n \geq 0$ such that

$$\prod_n E(k_n;\frac{z}{a_n})$$

converges uniformly in any given circle $|z| \leq R$, thus yielding an entire function of the desired nature. It is only necessary to choose k_n so as to insure convergence, for every R, of the series

$$\sum_n (\frac{R}{|a_n|})^{1+k_n}.$$

In particular, if for some <u>fixed</u> integer $k \geqq 0$ the sum

$$(4) \qquad \sum |a_n|^{-(1+k)}$$

converges, then the canonical product

$$(5) \qquad f_0(z) = \prod_n E(k; \frac{z}{a_n})$$

will solve Weierstrass's problem.

Once <u>one</u> function with the given zeros has been constructed, the most general solution is obtained by multiplying it by any zero-free entire function of the general type $e^{\phi(z)}$ ($\phi(z)$ entire). Hence it is obviously impossible to draw, from the distribution of the zeros alone, any conclusions about the growth of the function to the effect that it is limited by this or that function of $|z| = r$. Only for the canonical products can such a limitation of growth be expected. And indeed, Poincaré showed (1883) that in case (3) converges, the canonical product (2) grows less rapidly than $e^{|z|}$; more precisely, if α is any positive number, then

$$|f(z)| < e^{\alpha r} \qquad \text{for } |z| = r$$

from a certain r on. More generally, let us assume that there is a positive exponent λ lying between the integers k and k + 1, $k < \lambda \leqq k + 1$, such that $\sum |a_n|^{-\lambda}$ converges. One will then form the canonical product (5) of type k. Since, in forming this product, we have added to each term $1 - (z/a_n)$ a factor whose exponent is a polynomial of z of degree k, it seems natural to study the whole class of functions

$$(6) \qquad f(z) = e^{P(z)} \cdot f_0(z),$$

where P(z) is any polynomial of formal degree k. It is then true, as Poincaré proved, that for any $\alpha > 0$ and for sufficiently large $r > r(\alpha)$,

$$|f(z)| < e^{\alpha r^{\lambda}}$$

As long as we are only interested in an upper bound of $|f(z)|$ which depends on r, it seems reasonable to introduce the maximum M(r) of $|f(z)|$ on the circumference $|z| = r$. The last statement of the previous paragraph may then be expressed by the formula

$$\log M(r) = o(r^{\lambda}).$$

However, it is by no means to be expected that $|f(z)|$ is of a magnitude comparable to M(r) on the whole circumference $|z| = r$. This feature is peculiar to polynomials and must clearly break down in case there are infinitely many zeros; but even for a function without zeros like e^z the logarithm of its modulus oscillates between such wide margins as +r and -r along the periphery $|z| = r$.

We observe in passing that log M(r) is an increasing convex function of log r. To prove this, notice that the inequality $|f(z)| \leqq M(r)$ carries over from the periphery of the circle of radius r to the interior. Hence M(r) increases with r. Moreover set

$$\log r = u, \qquad \log M(r) = S$$

and let $r_1 < r < r_2$; then form the linear interpolation

$$S_1 + \frac{u-u_1}{u_2-u_1}(S_2-S_1) = H(u)$$

which satisfies $H(u_1) = S_1$, $H(u_2) = S_2$. On the two circles of radii r_1, r_2 we have

(7) $$\log |f(z)| - H(u) \leqq 0. \qquad (u = \log |z|).$$

The left side is a harmonic function in the annulus bounded by these two circles, with singularities of the type $-h \cdot \log (1/|z-a|)$, where the integer h is positive, so that the harmonic function becomes negatively infinite for $z \to a$. Consequently the inequality (7) holds everywhere in the annulus, in particular for $|z| = r$, and thus $S \leqq H(u)$ or

$$\begin{vmatrix} \log r_1, & \log r, & \log (r_2) \\ \log M(r_1), & \log M(r), & \log M(r_2) \\ 1, & 1, & 1 \end{vmatrix} \leqq 0$$

$$\text{for } r_1 < r < r_2.$$

It was Laguerre who first realized explicitly that one has to combine the distribution of the zeros with the degree of the polynomial $P(z)$, as our integer k does, in order to obtain a characteristic number for which one can expect interesting theorems. Laguerre and Borel use the terms "genus k" and "order λ" in conjunction with a function (6),

$$f(z) = e^{P(z)} \cdot \prod_n E(k; \frac{z}{a_n}),$$

for which $\sum |a_n|^{-\lambda}$ converges, $P(z)$ is a polynomial of formal degree k, and $k < \lambda \leqq k + 1$.* The converse of Poincaré's theorem on entire functions of finite genus is due to Hadamard (1893). It naturally consists of two parts; with some modifications and in a slightly sharpened form due to R. Nevanlinna, it states: For any $\lambda > 0$, convergence of the integral

$$\int^{\infty} \frac{\log M(r)}{r^{\lambda+1}} dr$$

implies convergence of the series $\sum |a_n|^{-\lambda}$. If $\log M(r) = o(r^{k+1})$ (k integral) then

---

* According to this definition, genus k and order λ are by no means uniquely determined by the function. In the literature it is more usual to modify (and complicate) the definition so as to bring about uniqueness.

$$f(z) = e^{P(z)} \cdot \lim_{R\to\infty} \prod_{|a_n| \leq R} E(k; \frac{z}{a_n})$$

where P(z) is a polynomial of formal degree k. The driving force for Hadamard's investigations was the wish to obtain sufficient information about the zeros of the ζ-function to establish the asymptotic law for the distribution of prime numbers. That law states that the number $\pi(n)$ of primes less than n becomes infinite with $n \to \infty$ exactly as strongly as n/log n:

$$\frac{\pi(n) \cdot \log n}{n} \to 1 \qquad \text{for } n \to \infty. \tag{8}$$

Riemann had shown how this prime number problem depends on the zeros of the ζ-function, and in 1896 both Hadamard himself and de la Vallée Poussin were able to draw the conclusion (8) from Hadamard's results concerning entire functions.

Besides the zeros, i. e. the solutions z of the equation f(z) = 0, one should study, for any preassigned value c, the c- places of f, i. e. the points z for which f(z) assumes the value c. We have seen that the 0-places might be spread much more thinly than the growth of the function leads one to expect (there might be even no zeros at all). But if this is so, then 0 is an exceptional value of f inasmuch as for almost all values $c \neq 0$ we can expect the distribution of c-places to conform with the growth. The first and most important step in this direction is Picard's famous theorem (1880) stating that an entire function assumes all values with the possible exception of only one. If one counts also $w = \infty$ as a value, one may say instead that a meromorphic function w = f(z) omits at most <u>two</u> values. For if it omits the value c, then 1/(f(z)-c) is an entire function. Picard's proof was simple, but made use of a highly transcendental tool, the modular function. By more elementary, yet much more complicated methods, E. Borel succeeded in bringing Picard's theorem into closer contact with the main body of the Poincaré-Hadamard theory of entire functions.

Two of the most penetrating investigators in the field of entire functions, along the lines opened by these great initiators, are Georges Valiron and Anders Wiman.

A long time passed by and the first World War shook our civilization before the basic paper on the theory of meromorphic functions was written by R. Nevanlinna in 1925. It put, at the same time, the more special theory of entire functions on a new foundation, greatly improving the older results. The appearance of this paper has been one of the few great mathematical events in our century. Rolf Nevanlinna's approach was greatly simplified by his brother Frithiof, and by L. Ahlfors; Ahlfors is also the author of a new method in this field, of half topological character. The preface tells how the theory of meromorphic curves came into being. A good part of the theory of meromorphic functions is better understood when looked upon as a special case of the theory of meromorphic curves.

One knows that an algebraic function y(x) (or an algebraic plane curve) is defined by an algebraic equation

$$y^m + f_1(x)y^{m-1} + \dots + f_m(x) = 0$$

whose coefficients are rational functions of x. After introduction of the corresponding Riemann surface $\mathfrak{R}$, both the independent and the dependent variables x and y appear as meromorphic functions on $\mathfrak{R}$. There is no reason why one should not simultaneously study three or four, or even n meromorphic functions on $\mathfrak{R}$ instead of two, thus passing from plane to n-dimensional algebraic curves. This will be our viewpoint. We shall spend a good deal of time on meromorphic curves, but afterwards we shall carry all the main results over to the general case of analytic curves where the z-plane is replaced by an arbitrary

Riemann surface.

I conclude this introduction by quoting the papers and books which form the chief landmarks of the historic development of our subject. (The numbers in front are used for reference; books are marked by an asterisk.)

[1] 1876 Weierstrass, Zur Theorie der eindeutigen analytischen Funktionen, Abhdlg. Ak. Wiss. Berlin, 1876, 11-60 = Math. Werke II, 77-124.

[2] 1880 E. Picard, Mémoire sur les fonctions entières, Ann. Ec. Norm. (2) 9, 147-166.

[3] 1883 H. Poincaré, Sur les fonctions entières, Bull. Soc. Math. de France 11, 136.

[4] 1893 J. Hadamard, Étude sur les propriétés des fonctions entières ..., Jour. de Math. (4) 9, 171-215.

[5] 1900 *E. Borel, Leçons sur les fonctions entières, Paris, 1900 (2e ed. 1921).

[6] 1903 *E. Borel, Leçons sur les fonctions méromorphes, Paris, 1903.

[7] 1925 R. Nevanlinna, Zur Theorie der meromorphen Funktionen, Act. Math. 46, 1925, 1.

[8] 1935 L. Ahlfors, Ueber eine Methode in der Theorie der meromorphen Funktionen, Soc. Sci. Fenn. Comm. phys.-math. 8, Nr. 10.

[9] 1935 L. Ahlfors, Zur Theorie der Ueberlagerungsflächen, Act. Math. 65, 157-194.

[10] 1936 *R. Nevanlinna, Eindeutige analytische Funktionen, Berlin, 1936.

[11] 1938 H. and J. Weyl, Meromorphic Curves, Ann. of Math. 39, 1938, 516-538.

[12] 1941 L. Ahlfors, The Theory of Meromorphic Curves, Acta Soc. Sci. Fenn., Ser. A, vol. 3, No. 4.

CHAPTER I

GEOMETRIC AND FUNCTION-THEORETIC FOUNDATIONS

A. PROJECTIVE SPACE AND UNITARY METRIC

§1. Plücker coordinates and the calculus of forms

In an n-dimensional vector space p ($\leq n$) linearly independent vectors $a_1, \ldots, a_p$ span a p-dimensional linear subspace or, as we shall briefly say, a p-spread $\{a_1, \ldots, a_p\}$. Plücker taught us how to characterize such p-spreads by coordinates. First we choose n independent vectors $e_1, \ldots, e_n$ as a basis or coordinate system in terms of which any vector x is represented by its coordinates $(x_1, \ldots, x_n)$,

$$x = x_1 e_1 + \ldots + x_n e_n.$$

If various vectors are distinguished by a subscript, the subscript characterizing the coordinate is written second. From the coordinates of p given vectors $a_1, \ldots, a_p$,

$$\begin{vmatrix} a_{11}, & \ldots, & a_{1n} \\ \cdot & \cdots & \cdot \\ a_{p1}, & \ldots, & a_{pn} \end{vmatrix}$$

we form the determinants of degree p,

$$(1.1) \qquad A(i_1 \ldots i_p) = \begin{vmatrix} a_{1i_1}, & \ldots, & a_{1i_p} \\ \cdot & \cdots & \cdot \\ a_{pi_1}, & \ldots, & a_{pi_p} \end{vmatrix}$$

These are the Plücker coordinates. If and only if they all vanish, the p vectors are linearly dependent. Each of the p indices $i_1, \ldots, i_p$ in $A(i_1 \ldots i_p)$ runs from 1 to n, and A is a skew-symmetric function of them. Spanning the same p-spread by another set $b_1, \ldots, b_p$ we obtain the same coordinates except for a common non-vanishing factor.

Indeed if

$$b_k = \sum_h \lambda_{kh} a_h \qquad (k,h=1,\ldots,p)$$

then

$$B(i_1 \ldots i_p) = \rho \cdot A(i_1 \ldots i_p), \quad \rho = \det(\lambda_{kh}).$$

In this sense the Plücker coordinates are homogeneous co-orinates of the spreads.

Any skew-symmetric function $A(i_1 \ldots i_p)$ of p indices i ranging over the values 1,...,n shall be called _a_ p-_ad_. If desirable, its "rank" p is indicated by a superscript, $A = A^p$. For $p > n$ there is no other p-ad than zero. For $p \leq n$ it suffices to know the values of the $\binom{n}{p}$ _ordered components_ $A(i_1 \ldots i_p)$ for which $i_1 < i_2 < \ldots < i_p$; they may be independently chosen. It is clear that one can _add_ two p-ads and _multiply_ a p-ad _by a number_. Hence the p-ads are the "vectors" of a $\binom{n}{p}$-dimensional vector space. We express the definition (1.1) of the Plücker coordinates by the formula

$$A = [a_1, \ldots, a_p].$$

The p-ads of this form are said to be _special_. The p-ad E of which all ordered components vanish except $E(12\ldots p) = 1$ is special, namely $= [e_1, \ldots, e_p]$. We shall not investigate here what relations between the components of A characterize the special A. Let us mention only the simplest non-trivial case $n = 4$, $p = 2$ where this condition reads

$$A(12)\,A(34) + A(31)\,A(24) + A(23)\,A(14) = 0.$$

If a p-spread is spanned by p vectors $a_1, \ldots, a_p$ and a q-spread by q vectors $b_1, \ldots, b_q$, we can form the spread spanned by $a_1, \ldots, a_p, b_1, \ldots, b_q$. In case the two spreads have no vector in common except zero, the resulting spread is of dimensionality $r = p + q$ and commonly called their direct sum. But we propose to use a _multiplication_ sign $\times$ for the representing polyads and hence write

(1.2) $[a_1,\ldots,a_p] \times [b_1,\ldots,b_q] = [a_1,\ldots,a_p,b_1,\ldots,b_q]$.

This multiplication is evidently associative. The law of commutativity holds in the modified form

$$B \times A = (-1)^{pq}(A \times B)$$

for the product of a p-ad A and a q-ad B. One can write the definition (1.2) in terms of the ordered components as follows:

$$(A \times B)(i_1,\ldots,i_r) = \sum \pm A(j_1 \ldots j_p) \cdot B(k_1,\ldots,k_q).$$

Here $i_1 < \ldots < i_r$, and the sum at the right extends alternatingly over all $\binom{r}{q}$ permutations

$$j_1,\ldots,j_p;\ k_1,\ldots,k_q \qquad (j_1<\ldots<j_p;\ k_1<\ldots<k_q)$$

of $i_1,\ldots,i_r$. Thus one convinces oneself that the components of the product are uniquely determined by the components of the two factors. The definition extends at once to arbitrary polyads. Then our multiplication of course is distributive with respect to addition. The resulting calculus of general polyads has recently proved of considerable importance in several applications; it is the algebraic part of E. Cartan's calculus of differential forms. Here our interest is concentrated on the special polyads representing spreads; by (1.2) the product of two special polyads is again special.

After defining multiplication, the p-ad $[a_1,a_2,\ldots,a_p]$ may now be written as the product $a_1 \times a_2 \times \ldots \times a_p$. Thus there is a redundance of notations and we should adopt either the cross or the square bracket. Following the classical works on Ausdehnungslehre by Grassmann and on electromagnetism by H. A. Lorentz, I choose the latter, and hence will now write [A,B], with or without comma, instead of $A \times B$. If $a_1,\ldots,a_p$ are linearly independent vectors, x belongs to the spread $\{a_1,\ldots,a_p\}$ if and only if $[x,a_1,\ldots,a_p] = 0$, i. e. if $[x,A] = 0$. This shows that the p-ad $A = [a_1,\ldots,a_p]$ determines uniquely the spread $\{a_1,\ldots,a_p\} = \{A\}$. The relation $[AB] = 0$ for two special polyads $A = [a_1\ldots a_p]$, $B = [b_1\ldots b_q]$ indicates that the vectors $a_1,\ldots,a_p$, $b_1,\ldots,b_q$ are linearly dependent. If $A \neq 0$, $B \neq 0$ and $p, q \geq 1$ this means that the two spreads $\{A\}$, $\{B\}$ have a non-vanishing vector in common (intersection).

Up to now we have used a fixed coordinate system $e_1,\ldots,e_n$. On replacing it by another $e_i' = (e_{i1}',\ldots,e_{in}')$,

$$x = \sum_i x_i' e_i',$$

the coordinates x_i undergo the non-singular transformation

$$x_j = \sum_i x_i' e_{ij}', \qquad x = \underline{\sigma} x', \tag{1.3}$$

while the components $X(i_1\ldots i_p)$ of the product $[x_1,\ldots,x_p]$ of p vectors $x_1,\ldots,x_p$ undergo the transformation

$$X(j_1\ldots j_p) = \sum_i X'(i_1\ldots i_p) e_{i_1 j_1}' \ldots e_{i_p j_p}'. \tag{1.4}$$

We denote the latter by $\underline{\sigma}_p$,

$$\underline{\sigma}_p [x_1',\ldots,x_p'] = [\underline{\sigma} x_1',\ldots,\underline{\sigma} x_p'],$$

and include this law of transformation in the notion of a p-ad. The operations described above are invariant in this sense. Instead of (1.4) one may write for the ordered components

$$X(j_1\ldots j_p) = {\sum_i}' X'(i_1\ldots i_p) \cdot \begin{vmatrix} e_{i_1 j_1}', \ldots, e_{i_1 j_p}' \\ \cdot \quad \cdot \quad \cdot \quad \cdot \quad \cdot \\ e_{i_p j_1}', \ldots, e_{i_p j_p}' \end{vmatrix} \tag{1.4'}$$

$$(i_1 < \ldots < i_p;\ j_1 < \ldots < j_p).$$

A sum over the indices $i_1,\dots,i_p$ of a p-ad will be indicated by $\sum_i$ if each index runs independently from 1 to n, by $\sum_i'$ if the restriction $i_1 < \dots < i_p$ is imposed.

In affine vector geometry all coordinate systems are treated as having equal rights.

A linear transformation $\underline{\sigma}$ (even if singular) may also be interpreted as a linear mapping; the role of x and x' is then exchanged. A linear mapping carrying e_i into e_i' will carry $x = \sum x_i e_i$ into the vector $x' = \sum x_i e_i' = \underline{\sigma} x$ with the coordinates $x_j' = \sum x_i e_{ij}'$. The mapping $\underline{\sigma}$ of the vector space into itself induces the mapping $\underline{\sigma}_p$ of the p-ads.

A 0-ad is simply a scalar, a number which is independent of the coordinate system. An n-ad $A = [a_1 \dots a_n]$ has only one ordered component $A(12\dots n)$ which is the determinant

$$V = \det(a_{ij}) = |a_1,\dots,a_n|. \tag{1.5}$$

This number V depends on the coordinate system, transforming by the equation

$$V = V' \cdot \det(e_{ij}')$$

if the coordinates are transformed by (1.3). We call a number of this type a volume factor, because (1.5) is the volume of the parallelotope spanned by $a_1,\dots,a_n$. If we limit ourselves to transformations $\underline{\sigma}$ of determinant 1 (unimodular transformations) then there is no need for distinguishing between scalar and volume factor.

The 0, 1,...,n-spreads of our vector space are the elements of a corresponding projective n-space; the 1-spreads are called points, the 2-spreads lines, etc. Until quite recently it was customary to consider an element of rank $p \geq 1$ as the set of all points contained in it; as such it is of dimensionality p - 1. But this interpretation, assigning a particular role to the 1-elements, does not work for the 0-element, absence of which from projective geometry in its customary formulation disturbs in more than one way the inherent symmetry of the situation. It is better to treat the elements of all ranks on equal footing. The affine vector space $\mathfrak{G}$ is the mother space from which the elements of the projective space are derived. When projective geometry is erected on an axiomatic basis, one should aim at constructing as directly as possible the number system and this affine mother space.

§2. Duality

A linear form depending on a variable vector x,

$$(\alpha x) = \alpha_1 x_1 + \dots + \alpha_n x_n,$$

may be characterized by its coefficients $(\alpha_1, \dots, \alpha_n)$ and thus looked upon as a vector in another n-dimensional vector space $\mathfrak{G}^*$, the dual space. If the x undergo the transformation (1.3) then the coefficients α_i are changed by the contragredient transformation

$$\alpha'_i = \sum_j e'_{ij}\alpha_j$$

such that

$$\sum_i \alpha'_i x'_i = \sum_i \alpha_i x_i.$$

This is the essential point in the relationship of the two vector spaces: that any change of coordinates in the one induces the contragredient change in the other. Therefore the product

$$(\xi x) = (\xi, x) = \sum \xi_i x_i$$

of a vector x in the original space and a vector ξ in the dual space has an invariant significance. The relation is mutual. Often one speaks of the vectors in $\mathfrak{G}$ and $\mathfrak{G}^*$ as covariant and contravariant vectors. We shall distinguish them by the Roman and Greek alphabets. Corresponding coordinate systems

$$e_1,\ldots,e_n \mid \epsilon_1,\ldots,\epsilon_n$$

in both spaces are connected by the relations

$$(e_i\epsilon_j) = \delta_{ij} = \begin{cases} 1 & (i=j) \\ 0 & (i\neq j). \end{cases}$$

From p vectors $\gamma_1,\ldots,\gamma_p$ in $\mathfrak{G}^*$ and p variable vectors $x_1,\ldots,x_p$ in $\mathfrak{G}$ we form the invariant

$$(2.1)\qquad \begin{vmatrix} (\gamma_1x_1),\ldots,(\gamma_1x_p) \\ \cdots\cdots\cdots \\ (\gamma_px_1),\ldots,(\gamma_px_p) \end{vmatrix} =$$

$$\sum_i \begin{vmatrix} \gamma_{1i_1},\ldots,\gamma_{1i_p} \\ \cdots\cdots \\ \gamma_{pi_1},\ldots,\gamma_{pi_p} \end{vmatrix} x_{1i_1}\cdots x_{pi_p}.$$

Hence it is possible to describe a contravariant p-ad Γ as a skew-symmetric multilinear form

$$\sum_i \Gamma(i_1,\ldots,i_p)x_{1i_1}\cdots x_{pi_p}$$

depending on p variable covariant vectors. In the same way a covariant p-ad is a skew multilinear form in p contravariant vectors. It is for this reason that Cartan has described our calculus of polyads as one of forms.

If we restrict ourselves to unimodular transformations (in accordance with Euler's original conception of affine geometry), then the determinant

$$|x_1,\ldots,x_p,\ c_{p+1},\ldots,c_n|$$

of p variable and n - p constant vectors x and c is likewise an invariant skew-linear form of $x_1,\ldots,x_p$. Thus we obtain a contravariant p-ad $\Gamma = {}^*C$ from a given covariant (n-p)-ad C by the relation

$$\Gamma(i_1,\dots,i_p) = \pm\, C(i_{p+1},\dots,i_n)$$

where the positive sign holds whenever $i_1,\dots,i_n$ is an even permutation. Clearly

$$C = (-1)^{p(n-p)}\cdot *\Gamma.$$

The extreme cases p = 0 and n are not excluded: the n-ad $C^n = [c_1\dots c_n]$ and the determinant $\Gamma^0 = |c_1,\dots,c_n|$ stand in the relation

$$C^n = *\Gamma^0, \qquad \Gamma^0 = *C^n.$$

In the sum at the right side of (2.1) one can introduce the restriction $i_\alpha \neq i_\beta$ for $\alpha \neq \beta$ and then write the sum as one extending over the ordered combinations $i_1 < \dots < i_p$ only:

$$\begin{vmatrix} (\gamma_1 x_1),\dots,(\gamma_1 x_p) \\ \cdot\ \cdot\ \cdot\ \cdot\ \cdot\ \cdot\ \cdot\ \cdot\ \cdot\ \cdot \\ (\gamma_p x_1),\dots,(\gamma_p x_p) \end{vmatrix} =$$

$$\sum_i{}' \Gamma(i_1\dots i_p)X(i_1\dots i_p) = (\Gamma X) = (X\Gamma)$$

for $\Gamma = [\gamma_1\dots\gamma_p]$ and $X = [x_1\dots x_p]$. The formula proves that the contravariant p-ads are transformed contragrediently to covariant p-ads. If

$$\Gamma^p = *C^{n-p}, \qquad X^p = *\Xi^{n-p},$$

then clearly

$$(\Gamma^p X^p) = (C^{n-p}\Xi^{n-p}).$$

The unit n-ad $E^n = [e_1\dots e_n]$ with the components

$$E^n(i_1\dots i_n) = \begin{cases} +1 \text{ if } (i_1\dots i_n) \text{ is an even permutation,} \\ -1 \text{ if } (i_1\dots i_n) \text{ is an odd permutation,} \end{cases}$$

is invariant with respect to unimodular transformations of the basis. Our definition of the star operation, $\Gamma^p = *C^{n-p}$, may now be written in the form

$$[X^p C^{n-p}] = (X^p \Gamma^p) \cdot E^n, \tag{2.2}$$

$X^p = [x_1 \ldots x_p]$ denoting a variable covariant p-ad.

If we drop the limitation to unimodular transformations, the * operation is no longer invariant, but the equation

$$V \cdot \Gamma^p = *C^{n-p}$$

establishes an invariant relation between a covariant (n-p)-ad C^{n-p}, a contravariant p-ad Γ^p and a volume factor V.

Any linear combination of $c_{p+1}, \ldots, c_n$ is a vector which "lies in" the (n-p)-spread $\{c_{p+1}, \ldots, c_n\} = \{C\}$. On the other hand, let us say that a vector ξ of $\mathfrak{G}^*$ "goes through" $\{C\}$ if satisfying the linear equations

$$(c_{p+1}\xi) = 0, \ \ldots \ , (c_n\xi) = 0. \tag{2.3}$$

According to the theory of linear equations, these ξ form a p-spread $\{\Gamma\}$ in $\mathfrak{G}^*$. Hence there is a one-to-one correspondence between the (n-p)-spreads in $\mathfrak{G}$ and the p-spreads in $\mathfrak{G}^*$, $\{C\} \rightleftarrows \{\Gamma\}$. This is the essence of the duality phenomenon in projective geometry. Its algebraic expression is the star operation. Indeed, assume $c_{p+1}, \ldots, c_n$ to be independent and extend them by p further vectors $c_1, \ldots, c_p$ so as to obtain a basis $c_1, \ldots, c_n$ for the full space $\mathfrak{G}$. Introduce the corresponding basis $\gamma_1, \ldots, \gamma_n$ in $\mathfrak{G}^*$:

$$(c_i \gamma_j) = \delta_{ij}.$$

From the multiplication theorem of determinants one gets at once

$$|x_1, \ldots, x_p, c_{p+1}, \ldots, c_n| \cdot |\gamma_1, \ldots, \gamma_n| = |(x_k \gamma_h)|.$$

The x are variable vectors in $\mathfrak{G}$ and the indices k, h run from 1 to p. In particular, for p = 0,

$$|c_1, \ldots, c_n| \cdot |\gamma_1, \ldots, \gamma_n| = 1.$$

Hence multiplication by the volume factor $V = |c_1, \ldots, c_n|$ yields the desired relation

$$V \cdot |(x_k \gamma_h)| = |x_1, \ldots, x_p, c_{p+1}, \ldots, c_n|$$

or

$$V \cdot [\gamma_1, \ldots, \gamma_p] = *[c_{p+1}, \ldots, c_n].$$

This construction proves that the * operation carries a special (n-p)-ad into a special p-ad.

§3. Unitary geometry

From affine, we pass to Euclidean metric geometry by introduction of the positive-definite quadratic form $x_1^2 + \ldots + x_n^2$. However, this form loses its positive character as soon as the coordinates are no longer limited to real values but become capable of arbitrary complex values, which will be the case in our applications. Then it must be replaced by the Hermitian unit form $x_1\bar{x}_1 + \ldots + x_n\bar{x}_n$. Starting from a definite coordinate system, we are thus led to introduce the scalar product $(x|y)$ of two vectors x,y by

$$(3.1) \qquad (x|y) = x_1\bar{y}_1 + \ldots + x_n\bar{y}_n.$$

At the same time we define the scalar product of two p-ads X, Y by

$$(3.2) \qquad (X|Y) = \sum_i{}' X(i_1 \ldots i_p)\bar{Y}(i_1 \ldots i_p).$$

The same expressions are used in the dual space.

Two vectors x, y are said to be perpendicular if $(x|y) = 0$. We ascribe to the vector x a length or magnitude $|x| \geq 0$ given by $|x|^2 = (x|x)$, and in the same way set $|X|^2 = (X|X)$. Geometrically $|[x_1,\ldots,x_p]|$ may be interpreted as the magnitude of the p-dimensional parallelotope spanned by $x_1,\ldots,x_p$. Notice that $|{*X}| = |X|$.

"Distances" in the projective space which depend on the ratios of coordinates only can be formed as follows. Distance of a point x and a plane ξ:

$$\|x\xi\| = \frac{|(x\xi)|}{|x|\cdot|\xi|}. \tag{3.3}$$

The Cauchy-Schwarz inequality

$$|u_1v_1+\ldots+u_mv_m|^2 \leqq (u_1\bar{u}_1+\ldots+u_m\bar{u}_m)(v_1\bar{v}_1+\ldots+v_m\bar{v}_m)$$

shows at once that this distance $\|x\xi\| \geqq 0$ and $\leqq 1$. It vanishes in case of incidence, $(x\xi) = 0$. In the same manner for p-elements:

$$\|X\Xi\| = \frac{|(X\Xi)|}{|X|\cdot|\Xi|}. \tag{3.4}$$

But all cases may be subsumed under the distance

$$|X:Y| = \frac{|[XY]|}{|X|\cdot|Y|} \tag{3.5}$$

of a p-element X and a q-element Y. This distance equals unity if p or q = 0. For $p \geqq 1$, $q \geqq 1$ its vanishing indicates intersection of $\{X\}$ and $\{Y\}$. The distance (3.4) arises from it by substitution of $*\Xi$ for Y ($q=n-p$). The relation

$$|X:Y| \leqq 1, \qquad |[XY]| \leqq |X|\cdot|Y| \tag{3.6}$$

does not seem to hold for general polyads X and Y, but we shall presently prove it in case X and Y are special.

In order to free these algebraic formulas from the absolute coordinate system on which they are based, we replace the explicit definition (3.1) of the scalar product by an <u>axiomatic description</u> which makes no reference to any special coordinates. Let us therefore say that our vector space is endowed with a metric if any two vectors x, y determine a scalar product $(x|y)$ which has the following four properties. It is linear in x,

(i) $(x+x'|y) = (x|y) + (x'|y), \qquad (\alpha x|y) = \alpha(x|y),$

"anti-linear" in y,

(ii) $(x|y+y') = (x|y) + (x|y'), \qquad (x|\alpha y) = \bar{\alpha}(x|y)$

(α being any number); it satisfies the Hermitian condition,

(iii) $(y|x)$ conjugate to $(x|y)$,

and is positive definite,

(iiii) $(x|x) > 0$ except for $x = 0$.

Because of the last property $(x|y)$ is non-degenerate, i. e. there is no vector a except 0 for which $(a|y) = 0$ identically in y.

In any coordinate system e_i, $x = x_1e_1 + \dots + x_ne_n$, the scalar product is expressible as a "Hermitian form"

$$G(x,y) = \sum g_{ij}x_i\bar{y}_j = \sum g(ij)x_i\bar{y}_j$$

the coefficients of which $g_{ij} = (e_i|e_j)$ satisfy the conditions $g_{ji} = \bar{g}_{ij}$. The form is positive definite, i. e.

$$G(x,x) > 0 \text{ except for } x = 0.$$

Hence we may simply say that a metric is introduced into the affine vector space by designating a positive definite Hermitian form as metric ground form.

We establish the existence of a basis $e_1,\dots,e_n$ in terms of which $(x|y)$ is expressed by the Hermitian unit form

$$E(x,y) = x_1\bar{y}_1 + \dots + x_n\bar{y}_n.$$

In other words, we are going to construct n vectors e_i such that

$$(3.7) \qquad (e_i | e_j) = \delta_{ij}.$$

Their linear independence is a consequence of these relations. Indeed, for $x = x_1 e_1 + \dots + x_n e_n$ they give

$$(x | e_j) = x_j;$$

hence the vanishing of x implies that of all x_j. The construction of the "normal" or "unitary" basis e_i follows the pattern of the construction by which one forms a Cartesian coordinate system in Euclidean geometry, by laying off the length 1 first in an arbitrary direction, then in a direction perpendicular to the first, then in a direction perpendicular to the first two etc. In more abstract terms: let us suppose we have succeeded in ascertaining $h - 1 < n$ vectors satisfying the relations (3.7) for i, j = 1,...,h-1. Then the h - 1 linear equations

$$(x | e_1) = 0, \dots, (x | e_{h-1}) = 0$$

have a non-vanishing solution x. A factor ρ can be so chosen that ρx is of "unit length":

$$(\rho x | \rho x) = \rho \bar{\rho} (x | x) = 1.$$

In this normalized form we call the solution e_h and then have

$$(e_h | e_1) = 0, \dots, (e_h | e_{h-1}) = 0, \quad (e_h | e_h) = 1$$

and also

$$(e_i | e_h) = \overline{(e_h | e_i)} = 0 \qquad \text{for } i=1,\dots,h-1.$$

The scalar product $(x|y)$ associates with any covariant vector y a linear form of x, i.e. a contravariant vector

$\eta = \underline{\mu} y$ so that

$$(x\eta) = (x|y).$$

The mapping $\underline{\mu}$ of $\mathfrak{G}$ into $\mathfrak{G}^*$ is anti-linear and non-singular, because $\underline{\mu}y$ never vanishes unless $y = 0$. Therefore it has an inverse $y = \underline{\mu}^{-1}\eta$, and the scalar product in the dual space may be defined by

$$(\xi|\eta) = (\xi,\underline{\mu}^{-1}\eta).$$

It is evidently linear in ξ, anti-linear in η, and the properties (iii) and (iiii), namely

$$(\eta,\underline{\mu}^{-1}\xi) = \overline{(\xi,\underline{\mu}^{-1}\eta)}, \quad (\xi,\underline{\mu}^{-1}\xi) > 0 \qquad \text{if } \xi \neq 0,$$

follow from the corresponding properties of $(x|y) = (x,\underline{\mu}y)$ by the substitution $\underline{\mu}^{-1}\xi = x$, $\underline{\mu}^{-1}\eta = y$.

In a normal coordinate system the relations $\eta = \underline{\mu}y$, $y = \underline{\mu}^{-1}\eta$ are expressed by

$$\eta_i = \overline{y}_i, \quad y_i = \overline{\eta}_i,$$

and

$$(\xi|\eta) = \xi_1\overline{\eta}_1 + \ldots + \xi_n\overline{\eta}_n.$$

In an arbitrary coordinate system in which

$$(x|y) = \sum g_{ij}x_i\overline{y}_j$$

we find for

$$\eta = \underline{\mu}y: \qquad \eta_i = \sum_j g_{ij}\overline{y}_j,$$

hence for

$$y = \underline{\mu}^{-1}\eta: \qquad y_i = \sum_j \gamma_{ij}\overline{\eta}_j$$

where the matrix $\|\gamma_{ij}\|$ is the conjugate of the inverse of $\|g_{ij}\|$. Thus

$$(\xi|\eta) = \sum \gamma_{ij}\xi_i\overline{\eta}_j.$$

As any linear mapping, so does the anti-linear mapping $\underline{\mu}$ induce a mapping $\underline{\mu}_p$ of the p-ads:

$$\Xi = \underline{\mu}_p X, \quad \underline{\mu}_p[x_1,\ldots,x_p] = [\underline{\mu}x_1,\ldots,\underline{\mu}x_p],$$

in coordinates

$$\Xi(i_1\ldots i_p) = \sum_j g(i_1 j_1)\ldots g(i_p j_p)\overline{X}(j_1\ldots j_p)$$

or in terms of ordered components

$$\Xi(i_1\ldots i_p) = \sum_j{}' \begin{vmatrix} g(i_1 j_1),\ldots,g(i_1 j_p) \\ \cdot\quad\cdot\quad\cdot\quad\cdot\quad\cdot\quad\cdot\quad\cdot \\ g(i_p j_1),\ldots,g(i_p j_p) \end{vmatrix} \cdot \overline{X}(j_1\ldots j_p).$$

We have the general rule

$$\underline{\mu}_{p+q}\,[A^p B^q] = [\underline{\mu}_p A^p, \underline{\mu}_q B^q].$$

The scalar product for p-ads is defined by

$$(X|Y) = (X, \underline{\mu}_p Y).$$

The fact that $(X|Y)$ satisfies the conditions (iii) and (iiii) is most easily verified by expressing it in terms of normal coordinates whereby we fall back on the old expression (3.2). For special p-ads $X = [x_1\ldots x_p]$, $Y = [y_1\ldots y_p]$ one finds

$$(X|Y) = \begin{vmatrix} (x_1|y_1),\ldots,(x_1|y_p) \\ \cdot\quad\cdot\quad\cdot\quad\cdot\quad\cdot\quad\cdot\quad\cdot \\ (x_p|y_1),\ldots,(x_p|y_p) \end{vmatrix}.$$

After defining the scalar product $(\Xi|H)$ of two contravariant p-ads Ξ, H in similar fashion, the relation

$$(\underline{\mu}_p X|\underline{\mu}_p Y) = (Y|X)$$

is obtained, in particular $|\underline{\mu}_p X| = |X|$.

We now see that all the formulas of the first part of this section hold in any normal coordinate system. As the transformations connecting them are called <u>unitary transformations</u>, we speak of <u>unitary geometry</u>.

Since the operation of passing to the conjugate complex commutes with *, we have in a normal coordinate system

$$\mu_{n-p}^{-1}(*A^p) = *\mu_p A^p.$$

However, since the operation * is not of unqualified invariance, one will not be surprised that in an arbitrary coordinate system this relation assumes the form

$$g\cdot\mu_{n-p}^{-1}*A^p = *\mu_p A^p \tag{3.8}$$

with the factor $g = \det(g_{ij})$. Let E^n, E^n be the unit n-ads in $\mathfrak{G}$ and $\mathfrak{G}^*$; then clearly $\mu_n E^n = g\cdot\mathrm{E}^n$ or $*\mu_n E^n = g$. In view of the definition of * this verification of (3.8) for $p = n$, $A^p = E^n$ suffices. Indeed, that definition (2.2),

$$[X^{n-p}, A^p] = (X^{n-p}, *A^p)\cdot E^n$$

when submitted to the fundamental metric operation μ_n turns into

$$[\Xi^{n-p}, \mu_p A^p] = \overline{(X^{n-p}, *A^p)}\cdot\mu_n E^n = (\Xi^{n-p}, \mu_{n-p}^{-1}*A^p)\cdot g\mathrm{E}^n$$

with $\Xi^{n-p} = \mu_{n-p}X^{n-p}$, and thus we find

$$(\Xi^{n-p}, *\mu_p A^p) = g\cdot(\Xi^{n-p}, \mu_{n-p}^{-1}*A^p).$$

Once (3.8) is proved, we obtain for any two p-ads X and A:

$$(X|A) = (X, \mu_p A) = (*X, *\mu_p A) = g\cdot(*X, \mu_{n-p}^{-1}*A) = g\cdot(*X|*A),$$

in particular

$$|X|^2 = g\cdot|*X|^2,$$

an equation which in a normal coordinate system reduces

to the previously mentioned $|X|^2 = |{*X}|^2$.

For later use we write down the formulas for a coordinate system in which $(x|y)$ assumes a form in which all coefficients g_{ij} off the principal diagonal, $i \neq j$, vanish

$$(x|y) = \kappa_1 x_1 \overline{y}_1 + \dots + \kappa_n x_n \overline{y}_n \qquad (\kappa_i \text{ real constants}),$$

$$(\xi|\eta) = \frac{1}{\kappa_1}\xi_1\overline{\eta}_1 + \dots + \frac{1}{\kappa_n}\xi_n\overline{\eta}_n, \tag{3.9}$$

$$(X|Y) = \sum_i{}' \kappa_{i_1} \dots \kappa_{i_p} X(i_1 \dots i_p)\overline{Y}(i_1 \dots i_p). \tag{3.10}$$

§4. Projection

Let an h-spread $\{E^h\}$ in $\mathfrak{G}$ be given. The space $\mathfrak{G}$ changes into an (n-h)-dimensional vector space $\mathfrak{G}/E^h$ by identifying any two vectors x, y which are congruent modulo $\{E^h\}$, i.e. the difference of which lies in $\{E^h\}$. In affine language, this process is described as <u>parallel projection along</u> $\{E^h\}$; in projective language, as <u>central projection from</u> $\{E^h\}$. A linear form (αx) is a form in $\mathfrak{G}/E^h$ if it vanishes for all vectors x in $\{E^h\}$; hence the dual of $\mathfrak{G}/E^h$ is a sub-space $\{E^{n-h}\}$ of $\mathfrak{G}^*$. Let $\{E^h\}$ be spanned by the first h unit vectors $e_1, \dots, e_h$. Then the vectors in $\mathfrak{G}/E^h$ have $x_{h+1}, \dots, x_n$ as their coordinates because two vectors x, y are congruent modulo $\{E^h\}$ if and only if their last $n - h$ coordinates coincide; whereas the vectors in $\{E^{n-h}\}$ are the vectors of the form $0, \dots, 0, \xi_{h+1}, \dots, \xi_n$.

We wish to describe how <u>multiplication by E^h represents projection</u>. Indeed, for any $X = [x_1 \dots x_p]$ the product $[E^h X] = [E^h x_1 \dots x_p]$ does not change if each x is replaced by a vector congruent to x modulo $\{E^h\}$; hence it is a quantity in $\mathfrak{G}/E^h$. Let E^h again be spanned by $e_1, \dots, e_h$, $E^h = [e_1 \dots e_h]$. One finds at once that any ordered component $Y(i_1 \dots i_h j_1 \dots j_p)$ of $Y = [E^h X]$ vanishes unless $i_1 = 1, \dots, i_h = h$, whereas for the excluded combinations, in which the j's range between h + 1 and n,

$$Y(1\ldots hj_1\ldots j_p) = X(j_1\ldots j_p).$$

In this precise sense $Y = [E^hX]$ may be identified with the projection $\tilde{X}$ of X which, with indices j ranging from h + 1 to n, is defined by

$$\tilde{X}(j_1\ldots j_p) = X(j_1\ldots j_p).$$

According to our definition of a metric space, any h-spread $\{E^h\}$ in the metric space $\mathfrak{G}$ is itself an h-dimensional metric space, and hence we can refer it to a normal coordinate system $e_1,\ldots,e_h$. By the same construction we can extend this basis of $\{E^h\}$ to a normal coordinate system $e_1,\ldots,e_h, e_{h+1},\ldots,e_n$ for the full space $\mathfrak{G}$. Then $e_{h+1},\ldots,e_n$ span the space $\{\sim E^h\}$ perpendicular to $\{E^h\}$. Projection along $\{E^h\}$ is naturally interpreted as projection upon $\{\sim E^h\}$, every vector x being congruent modulo $\{E^h\}$ to a uniquely determined vector $\tilde{x}$ in $\{\sim E^h\}$. If we set $E^h = [e_1\ldots e_h]$ and use $e_1,\ldots,e_n$ as coordinate system, then the "identification" of the projection $\tilde{X}$ with $[E^hX]$ described above yields the equations

$$|E^h| = 1, \qquad |[E^hX]| = |\tilde{X}|$$

and the inequality $|\tilde{X}| \leqq |X|$. In the latter, the equality sign prevails if and only if $X = [x_1\ldots x_p]$ is spanned by vectors $x_1,\ldots,x_p$ in $\{\sim E^h\}$. Drop the normalization $|E^h| = 1$ and in order to avoid confusion with the unit h-ad write $A = A^h$ instead of E^h. Then

$$|\tilde{X}| = \frac{|[AX]|}{|A|}$$

is the magnitude of the projection of X along $\{A\}$ upon $\{\sim A\}$. Thus we have proved the inequality (3.6), $\|A:B\| \leqq 1$ for special p-ads A, B, obtaining the further result that, for $A \neq 0$, $B \neq 0$, the inequality becomes an equality if and only if $\{A\}$ and $\{B\}$ are perpendicular. In affine language $\|A:X\|$ may be interpreted as the ratio at which the magnitude of X shrinks by projection along $\{A\}$. We have the obvious <u>law of symmetry</u>:

> B shrinks by projection upon $\{\sim A\}$ at the same scale as A does by projection upon $\{\sim B\}$.

In projective language this fact is nothing but the symmetry of the distance $\|A:B\|$. Moreover we prove

LEMMA 4.A.

$$\{A^{h-1}\} \subset \{A^h\}$$

implies

$$\|A^{h-1}:X\| \geqq \|A^h:X\|. \tag{4.1}$$

This fact is pretty obvious from both the affine and the projective standpoints. We can ascertain a vector a perpendicular to $\{A^{h-1}\}$ such that $\{A^h\} = \{a, A^{h-1}\}$. Since $\{\sim A^h\} \subset \{\sim A^{h-1}\}$, projection upon $\{\sim A^h\}$ can be accomplished in two steps: first project upon $\{\sim A^{h-1}\}$ and then within that subspace along a. Our inequality states that the second process effects a further shrinkage. From the projective standpoint one expects that, e.g. the distance between a point and a line is smaller than or equal to the distance of that point from any point on the line. The formal proof goes as follows: Set $Y = [A^{h-1}X]$ so that $[A^hX] = [aY]$. Then

$$|[aY]| \leqq |a| \cdot |Y|, \tag{4.2}$$

and because a is perpendicular to $\{A^{h-1}\}$,

$$|A^h| = |a| \cdot |A^{h-1}|. \tag{4.3}$$

Divide (4.2) by (4.3) and thus obtain (4.1).

Let us for a moment return to the previous notation referring to a normal basis $e_1, \ldots, e_n$. Every vector x may be split into a component $\hat{x}$ in $\{E^h\}$ and one $\tilde{x}$ perpendicular to $\{E^h\}$, i.e. in $\{\sim E^h\}$. Thus a p-ad $X = [x_1 \ldots x_p]$ gives rise not only to the projection $\tilde{X} = [\tilde{x}_1 \ldots \tilde{x}_p]$, but also to the perpendicular $\hat{X} = [\hat{x}_1 \ldots \hat{x}_p]$. Clearly

$$(4.4) \qquad |\tilde{X}|^2 + |\hat{X}|^2 \leqq |X|^2,$$

because the left side is the square sum of all $X(i_1 \ldots i_p)$ in which the ordered combination $(i_1, \ldots, i_p)$ lies entirely in either the section $(1, \ldots, h)$ or the complementary section $(h+1, \ldots, n)$. For $p = 1$ the equality sign prevails. It is not necessary to introduce the perpendicular as a new notion besides projection, because the perpendicular along $\{A\}$ upon $\{\sim A\}$ is the projection along $\{\sim A\}$ upon $\{A\}$. Hence the magnitudes of perpendicular and projection upon $\{\sim A\}$ are given by $\frac{|[\sim A, X]|}{|\sim A|}$ and $\frac{|[A, X]|}{|A|}$, and the relation (4.4) turns into

$$\|\sim A : X^p\|^2 + \|A : X^p\|^2 \leqq 1 \qquad \text{(equality sign for } p = 1\text{)}.$$

The dual $\{*A\} = \{\beta_1, \ldots, \beta_{n-h}\}$ of the h-spread $\{A\} = \{a_1, \ldots, a_h\}$ is spanned by the $n - h$ independent solutions $\xi = \beta_k$ of the equations

$$(a_i \xi) = 0 \qquad (i=1, \ldots, h).$$

Write $\beta_k = \underline{\mu} b_k$. Then one sees that

$$(4.5) \qquad \{\sim A\} = \underline{\mu}_{n-h}^{-1} {*A^h}$$

or by a similar argument

$$\{\sim A\} = \{*\underline{\mu}_h A^h\}.$$

The identity of these two elements has previously been established under (3.8). Because of (4.5) the magnitude of the perpendicular

$$(4.6) \qquad \frac{|[\sim A, X^p]|}{|\sim A|} \quad \text{equals} \quad \frac{|[*A, \underline{\mu}_p X^p]|}{|*A|}$$

§5. Distortion

In our investigations unitary metric will play the part of an auxiliary construction. Hence we should know to what extent the quantities with which we deal are influenced by the choice of the metric. Let us therefore compare two metrics defined by two positive definite Hermitian forms

$$(x|y)_o = G_o(x,y) = \sum g^o_{ij} x_i \bar{y}_j, \quad (x|y) = G(x,y) = \sum g_{ij} x_i \bar{y}_j.$$

For the "distortion" $\sigma = |x|/|x|_o$ of the length of an arbitrary vector we obtain the equation

$$\sigma^2 = G(x,x)/G_o(x,x).$$

If we look upon the metric denoted by the subscript o as the "true one", the determination of the extreme values of σ is nothing else than the transformation on to principal axes of the "ellipsoid" $G(x,x) = 1$ in a space where the "spheres" are given by $G_o(x,x) = 1$. This is the classical theorem about principal axes:

> Let G_o, G be two given Hermitian forms of which the first is positive definite. One then can find a coordinate system $e_1,\dots,e_n$ in which
>
> $$G_o(x,y) = x_1\bar{y}_1 + \dots + x_n\bar{y}_n,$$
> $$G(x,y) = \kappa_1 x_1\bar{y}_1 + \dots + \kappa_n x_n\bar{y}_n.$$
>
> The "eigen values" $\kappa_1,\dots,\kappa_n$ are real numbers.

The proof depends on the linear mapping $x \longrightarrow x'$ defined by the relation

$$G_o(x',y) = G(x,y)$$

holding identically in y, or in terms of coordinates by the relations

$$\sum_i x'_i g^0_{ij} = \sum_i x_i g_{ij}.$$

Because of the non-degeneracy of G_0 they are solvable with respect to x'_i. We seek a non-vanishing vector x which this mapping carries into a multiple κx of itself. We obtain the equations

$$(5.1) \qquad \sum_i x_i (\kappa g^0_{ij} - g_{ij}) = 0.$$

Hence we choose $t = \kappa$ as a root of the secular equation of degree n,

$$\det (t g^0_{ij} - g_{ij}) = 0,$$

and then find a non-vanishing solution $x = e$ of the set of homogeneous linear equations (5.1), which we normalize by the condition $G_0(e,e) = 1$. We then have

$$(5.2) \qquad G(e,y) = \kappa \cdot G_0(e,y)$$

identically in y, in particular

$$G_0(e,e) = 1, \qquad G(e,e) = \kappa .$$

This shows κ to be real. Write $\kappa = \kappa_1$, $e = e_1$ and split every vector x into

$$x = x_1 e_1 + x',$$

where x' is perpendicular to e_1 in the sense of the metric G_0, $G_0(x',e_1) = 0$. This is accomplished by taking $x_1 = G_0(x,e_1)$. The vectors x' satisfying $G_0(x',e_1) = 0$ form a subspace $\mathfrak{G}'$ of dimensionality $n - 1$. From (5.2) and the Hermitian nature of G_0 and G we obtain

$$\left.\begin{array}{l} G_0(e_1,y') = 0, \ G(e_1,y') = 0, \\ G_0(x',e_1) = 0, \ G(x',e_1) = 0 \end{array}\right\} \text{ for } x',y' \in \mathfrak{G}'.$$

Thus

469156

$$G_o(x,y) = x_1\overline{y}_1 + G_o(x',y'),$$
$$G(x,y) = \kappa_1 x_1\overline{y}_1 + G(x',y'),$$

and the problem has been reduced from n to n-1 dimensions.

In our case G is positive definite, and hence the eigen values κ are necessarily positive. We arrange them in ascending order,

$$\kappa_1 \leqq \kappa_2 \leqq \cdots \leqq \kappa_n.$$

The equations

$$(x|x)_o = |x_1|^2 + \cdots + |x_n|^2,$$
$$(x|x) = \kappa_1|x_1|^2 + \cdots + \kappa_n|x_n|^2$$

clearly imply the inequality

$$\kappa_1 \cdot (x|x)_o \leqq (x|x) \leqq \kappa_n \cdot (x|x)_o.$$

The lower and upper bounds are actually assumed for $x = e_1$ and $x = e_n$ respectively. Hence the square of the distortion σ^2 for the lengths $|x|$ of covariant vectors x lies between the positive limits κ_1 and κ_n.

For contravariant vectors we find from (3.9):

$$\frac{1}{\kappa_n} \cdot (\xi|\xi)_o \leqq (\xi|\xi) \leqq \frac{1}{\kappa_1} \cdot (\xi|\xi)_o.$$

The corresponding limits are here $1/\kappa_n$ and $1/\kappa_1$.

Comparison of the formulas for p-ads,

$$(X|X)_o = \sum{}' |X(i_1 \dots i_p)|^2,$$
$$(3.10) \qquad (X|X) = \sum{}' \kappa_{i_1} \cdots \kappa_{i_p} |X(i_1 \dots i_p)|^2$$

shows that

$$m_p \cdot (X|X)_o \leqq (X|X) \leqq M_p \cdot (X|X)_o$$

where

$$m_p = \kappa_1 \ldots \kappa_p, \qquad M_p = \kappa_{n-p+1} \ldots \kappa_n.$$

The limits are assumed for $X = [e_1 \ldots e_p]$ and $[e_{n-p+1}, \ldots, e_n]$, respectively.

From the affine magnitudes we pass to the projective distances. For the square of the distortion of the point-plane distance (3.3) we get at once the limits

$$\kappa_1/\kappa_n \text{ and } \kappa_n/\kappa_1,$$

and for the distance $\|X\Xi\|$ of a covariant and a contravariant p-element the result is that σ lies between λ_p and λ_p^{-1} where λ_p is defined by

$$\lambda_p^2 = M_p/m_p.$$

It is a little more difficult to ascertain the exact bounds for the distortion σ of the distance $\|X:Y\|$, (3.5), of a p-element X and a q-element Y ($r = p+q \leqq n$). The limits which can be fixed at once,

$$\frac{m_{p+q}}{M_p M_q} \leqq \sigma^2 \leqq \frac{M_{p+q}}{m_p m_q},$$

are far from being the sharpest. We use the last inequality only in the case $p + q = n$ where it gives as limits $\lambda_p = \lambda_q$ and its reciprocal. This is no wonder because we are then dealing with the same thing as $\|X\Xi\|$. But let us state this result explicitly as a lemma:

LEMMA 5.A. In the case $p + q = n$ the distortion of $\|X:Y\|$ lies between λ_p and λ_p^{-1} where λ_p^2 is the ratio of the maximum M_p and the minimum m_p of $|[x_1, \ldots, x_p]|^2$ for variable vectors $x_1, \ldots, x_p$ subject only to the restriction $|[x_1, \ldots, x_p]|_0^2 = 1$.

Comparison of the distance $\|X:Y\|$ in our two metrics is possible only if $[X,Y] \neq 0$, and then $\{X,Y\}$ is an r-dimensional space $\{E^r\}$ in which both $\{X\}$ and $\{Y\}$ are imbedded. Hence we determine first the limits $\lambda_p(E^r)$ and $\lambda_p^{-1}(E^r)$ of the distortion of $\|X:Y\|$ under the assumption that $\{X\}$ and $\{Y\}$ vary over the p-spreads and q-spreads in a given (p+q)-dimensional subspace $\{E^r\}$. An upper bound of $\lambda_p(E^r)$ which is independent of the subspace $\{E^r\}$ will then solve our problem. The first part is settled by our lemma: $\lambda_p^2(E^r)$ is the ratio between the maximum $M_p(E^r)$ and the minimum $m_p(E^r)$ of $|[x_1,\dots,x_p]|^2$ under the condition that $|[x_1,\dots,x_p]|_0^2 = 1$ <u>and that</u> $x_1,\dots,x_p$ <u>lie in</u> $\{E^r\}$. However, the underlined additional restriction lowers the maximum and raises the minimum.. Consequently

$$M_p(E^r) \leqq M_p, \qquad m_p(E^r) \geqq m_p$$

and thus $\lambda_p(E^r) \leqq \lambda_p$. We have now proved that the distortion for the distance of X and Y lies between λ_p and λ_p^{-1}. For the same reason it lies between λ_q and λ_q^{-1}. In view of the inequality $p + q \leqq n$ the better result is the one corresponding to the smaller of the two ranks p and q. Let us therefore assume $p \leqq q$ and then show by a definite example that the bounds λ_p and λ_p^{-1} may not be improved.

By means of a variable number δ which we shall let approach zero, we set

$$Y = E = [e_1,\dots,e_q], \; X = A(\delta) = [e_1+\delta e_{n-p+1},\dots,e_p+\delta e_n].$$

Because of $p \leqq q$,

$$\{E,A(\delta)\} = \{e_1,\dots,e_q, e_{n-p+1},\dots,e_n\}.$$

The square distortion of $|E|$ is $\kappa_1\dots\kappa_q$, the square distortion of $|A(\delta)|$ approaches $\kappa_1\dots\kappa_p$, whereas the square distortion of $|[E,A(\delta)]|$ is $\kappa_1\dots\kappa_q\cdot\kappa_{n-p+1}\dots\kappa_n$. Hence the square distortion of $\|A(\delta):E\|$ approaches

$$\kappa_{n-p+1}\dots\kappa_n/\kappa_1\dots\kappa_p = \lambda_p^2$$

as δ tends to zero. The bounds λ_p and λ_p^{-1}, although the best, will in general not be actually assumed in a non-degenerate fashion (i.e. for $[X,Y] \neq 0$).

THEOREM. Let two metrics be defined by two positive definite Hermitian forms with the coefficient matrices G, G_o. Arrange in ascending order the n roots $\kappa_1,\ldots,\kappa_n$ of the polynomial

$$\det (tG_o-G) = \det G_o \cdot (t-\kappa_1) \ldots (t-\kappa_n)$$

(they are all positive) and set

$$\lambda_h^2 = \kappa_{n-h+1} \ldots \kappa_n/\kappa_1 \ldots \kappa_h.$$

Let $p + q \leq n$ and $h = \min (p,q)$. The ratio of the projective distances $\|X:Y\|$ of a p-element X and a q-element Y computed on the basis of the two metrics, lies between λ_h and λ_h^{-1}. These bounds cannot be improved.

B. GENERALITIES CONCERNING ANALYTIC CURVES

§6. Analytic branch. Associated curves

A branch of an analytic curve $\mathfrak{C}$ is defined if the coordinates x_i of a variable point (1-element) are expressed as power series in terms of a complex parameter t,

$$x_i = \mathfrak{P}_i^*(t) \qquad (i=1,\ldots,n).$$

We suppose that the $\mathfrak{P}_i^*(t)$ converge in a circle $|t| < r$ of positive radius r, but do not all vanish identically. We may remove a common factor t^d from all $\mathfrak{P}_i^*(t)$,

$$\mathfrak{P}_i^*(t) = t^d \cdot \mathfrak{P}_i(t), \qquad \mathfrak{P}_i(t) = c_i + c_i't + c_i''t^2 + \ldots ,$$

such that $(c_1,\ldots,c_n) \neq (0,\ldots,0)$. Because the x_i are homogeneous coordinates we may then just as well set

$$x_i = \mathfrak{P}_i(t) \qquad (i=1,\ldots,n). \tag{6.1}$$

This "reduced" representation ascribes a definite position $c_1 : \ldots : c_n$ to the point $t = 0$ of the curve. Any reduced representation of the same branch arises from (6.1) by multiplication with a common "gauge factor" $\rho(t)$,

$$\rho(t) = \rho_0 + \rho_1 t + \ldots ,$$

which is a power series with non-vanishing constant term ρ_0 and converges in a circle of positive radius. (We do not require that it converge in the circle $|t| < r$; in other words we do not fix the neighborhood of $t = 0$ for which the representation is valid.) Only such properties of the branch have objective significance as are not affected by the arbitrary gauge factor ρ.

Let us say that the reduced representation (6.1) is valid in the circle K_r, $|t| < r$, if 1) the power series $\mathfrak{P}_i(t)$ converge in K_r and if 2) for no value t within K_r all n power series $\mathfrak{P}_i(t)$ vanish simultaneously. Let t_1 be any point in K_r. We make the substitution $t \to t_1 + t$ and develop $\mathfrak{P}_i(t_1+t)$ into a power series of t, $\mathfrak{P}_i(t_1+t) = \mathfrak{P}_i^1(t)$. The reduced representation $x_i = \mathfrak{P}_i^1(t)$ will be valid in a certain circle $|t| < r_1$ and constitutes the immediate analytic continuation of the branch with which we started. Had we started with the representation $x_i = \rho(t)\cdot\mathfrak{P}_i(t)$ differing from the first by a gauge factor $\rho(t)$ we should have obtained an analytic continuation $x_i = \rho_1(t)\cdot\mathfrak{P}_i^1(t)$ with a certain gauge factor $\rho_1(t)$ which arises from $\rho(t)$ by the same substitution $t \to t_1 + t$; this is true at least if $|t_1|$ is sufficiently small.

Assume now the embedding space of our curve to be the projective n-space. Then any coordinate system y_i arising from x_i by a non-singular linear transformation

$$x_j = \sum_i e_{ij}^! y_i \tag{6.2}$$

with constant coefficients $e_{ij}^!$ is equally admissible and we are thus led to focus our attention on such properties of the branch as remain unaffected by an arbitrary transformation of coordinates.

Up to now we have referred the curve to a definite parameter. However we come closer to the geometric concept of a curve by acting on the assumption that the choice of the parameter is arbitrary and not determined by the curve itself. Instead of t we may use as parameter any variable s which is connected with t by an analytic substitution

$$t = b_1 s + b_2 s^2 + \dots \qquad (b_1 \neq 0). \tag{6.3}$$

The power series on the right is subject only to the conditions (1) that it converge in a circle of positive radius ($|s| < q$), (2) that the initial coefficient b_1, the value of the derivative dt/ds at $s = 0$, does not vanish. The parameter transformations of this type form a group. Substitution of the power series (6.3) for t carries the representation $x_i = \mathfrak{P}_i(t)$ into an equivalent representation $x_i = \mathfrak{Q}_i(s)$ of the curve. Hence we are led to postulate invariance with respect to parameter transformations of type (6.3).

In the course of our discussion we have thus encountered <u>three types of invariance</u>: 1) with respect to a change of the gauge factor, 2) with respect to a change of the projective coordinate system, 3) with respect to a change of the parameter t in terms of which the branch is represented. As an example of a quantity which is invariant in all three respects, consider the multiplicity ν of the intersection of the branch at $t = 0$ with an arbitrary plane (α),

$$(\alpha x) = \sum_i \alpha_i x_i = 0, \qquad (\alpha_1, \dots, \alpha_n) \neq (0, \dots, 0).$$

If $\sum \alpha_i \mathfrak{P}_i(t)$ does not vanish identically, it vanishes for $t = 0$ with a certain order $\nu \geq 0$. It is essential that we use a <u>reduced</u> representation $x_i = \mathfrak{P}_i(t)$ for the determination of the multiplicity ν. Then it is evident that ν is affected neither by the gauge factor $\rho(t)$ (because of $\rho_0 \neq 0$) nor by the transformation (6.3) of the parameter t (because of $b_1 \neq 0$). In passing to another coordinate system, but keeping the plane (α) fixed, one has of course to submit the coefficients α_i to the contragredient transformation; in this sense the multiplicity ν of intersection is also independent of the coordinate system.

In case $\sum \alpha_i \mathfrak{p}_i(t)$ vanishes identically for some set of coefficients $(\alpha_1, \dots, \alpha_n) \neq (0, \dots, 0)$, then the curve lies in the plane (α). Under these circumstances the $\mathfrak{p}_i(t)$ are said to be linearly dependent and the curve to be <u>degenerate</u>. In terms of an appropriate system of coordinates, a number of the $\mathfrak{p}(t)$, say $\mathfrak{p}_{m+1}(t), \dots, \mathfrak{p}_n(t)$, will vanish identically, whereas $\mathfrak{p}_1(t), \dots, \mathfrak{p}_m(t)$ are linearly independent. The curve then lies in the m-subspace $x_{m+1} = \dots = x_n = 0$ but not in a subspace of lower dimensionality. The non-degenerate case is characterized by $m = n$. From the beginning we have excluded the case of <u>total</u> <u>degeneracy</u>, $m = 0$, where all $\mathfrak{p}_i(t) = 0$.

With our curve $\mathfrak{C} = \mathfrak{C}_1$, which is a one-parameter family of <u>points</u> or 1-elements, there is associated the family or "curve" $\mathfrak{C}_2$ of its <u>tangents</u> or osculating 2-elements, moreover the family $\mathfrak{C}_3$ of its osculating 3-elements etc. Let the accent, when affixed to the coordinates x_i of the curve, denote differentiation with respect to t. The Plücker coordinates of the tangent at the point t are the two-row determinants

$$X^2(i_1 i_2) = x_{i_1}(t) \cdot x'_{i_2}(t) - x_{i_2}(t) \cdot x'_{i_1}(t),$$

i.e. the components of the dyad $X^2 = [x, x']$. In the same manner the coordinates of the osculating p-element are the components of the p-ad $X^p = [x, x', \dots, x^{(p-1)}]$. The index p is capable of the values 0, 1, 2, ... , n. The highest, X^n, equals *W where W is the so-called Wronskian of the n functions $x_1(t),\dots,x_n(t)$,

$$W = \begin{vmatrix} x_1, & \dots, & x_n \\ x_1', & \dots & x_n' \\ \cdot & \cdots & \cdot \\ x_1^{(n-1)}, & \dots, & x_n^{(n-1)} \end{vmatrix}.$$

If the coordinates x_i are subjected to a linear transformation (6.2) with constant coefficients, then the derivatives x_i' undergo the same transformation and so do the higher derivatives x_i'', x_i''', etc. Hence our polyads X^1, X^2, ... are nothing but the [ ]-products of the vectors x, x', x", ... This remark settles the question of their behavior under arbitrary coordinate transformations. Less obvious is the fact that the element $\{X^p\}$ is independent of the gauge factor $\rho(t)$. However, on replacing $x_i(t)$ by $y_i(t) = \rho(t)\cdot x_i(t)$ we find

$$y_i'(t) = \rho(t)x_i'(t) + \rho'(t)x_i(t),$$
$$y_i''(t) = \rho(t)x_i''(t) + 2\rho'(t)x_i'(t) + \rho''(t)x_i(t),$$
$$\cdot \quad \cdot \quad \cdot \quad \cdot \quad \cdot \quad \cdot \quad \cdot \quad \cdot \quad \cdot \quad \cdot \quad \cdot \quad \cdot$$

and then readily

$$\begin{vmatrix} y_{i_1}, \dots, & y_{i_p} \\ y_{i_1}', \dots, & y_{i_p}' \\ \cdot \quad \cdot & \cdot \quad \cdot \\ y_{i_1}^{(p-1)}, \dots, & y_{i_p}^{(p-1)} \end{vmatrix} = \rho^p \begin{vmatrix} x_{i_1}, \dots, & x_{i_p} \\ x_{i_1}', \dots, & x_{i_p}' \\ \cdot \quad \cdot & \cdot \quad \cdot \\ x_{i_1}^{(p-1)}, \dots, & x_{i_p}^{(p-1)} \end{vmatrix},$$

so that all components $X^p(i_1,\dots,i_p)$ are simply multiplied by the same factor $\rho^p \neq 0$.

In the third place we have to ascertain how X^p is influenced by a change of parameter (6.3). The derivatives with respect to t and to s are connected by the relations

$$\frac{dx_i}{ds} = \frac{dx_i}{dt}\frac{dt}{ds},$$

$$\frac{d^2x_i}{ds^2} = \frac{d^2x_i}{dt^2}\left(\frac{dt}{ds}\right)^2 + \frac{dx_i}{dt}\frac{d^2t}{ds^2},$$

in general

$$\frac{d^px_i}{ds^p} = \frac{d^px_i}{dt^p}u_o^p + \frac{d^{p-1}x_i}{dt^{p-1}}u_1^p + \dots + \frac{dx_i}{dt}u_{p-1}^p$$

where the u satisfy the recursion formulas

$$u_o^{p+1} = u_o^p\frac{dt}{ds}, \qquad u_1^{p+1} = \frac{du_o^p}{ds} + u_1^p\frac{dt}{ds},$$

$$u_2^{p+1} = \frac{du_1^p}{ds} + u_2^p\frac{dt}{ds}, \dots, u_p^{p+1} = \frac{du_{p-1}^p}{ds}.$$

Therefore

$$u_o^p = \left(\frac{dt}{ds}\right)^p,$$

and the effect of replacing the parameter t by s amounts to multiplication of all components $X^p(i_1,\dots,i_p)$ by the

$$0 + 1 + \dots + (p-1) = \tfrac{1}{2}p(p-1)$$

power of dt/ds, so that the element $\{X^p\}$ itself, which is described by the ratio of these components, remains unaltered.

The questions as to the reduced representation as well as the possible degeneracy of $\mathfrak{C}_p$ -- for it may well happen that $\mathfrak{C}_p$ is totally degenerate, $X^p = 0$, even if $\mathfrak{C}$ is not -- are best settled by introducing a certain standard coordinate system. Of strictly local significance and putting into evidence the type of singularity that the curve $\mathfrak{C}$ possesses at $t = 0$, it resembles the fundamental trihedral at a point of a curve employed in metric differential geometry.

§7. Stationarity indices

THEOREM. Let us assume that the curve $\mathfrak{C}$ is non-degenerate, and let $x_i = x_i(t)$ be a reduced representation of $\mathfrak{C}$ by means of power series $x_i(t)$. We maintain that in an appropriately "adjusted" coordinate system the initial terms of these power series are as follows, formula (7.1):

$$x_1 = t^{\delta_1} + \dots, \quad x_2 = t^{\delta_2} + \dots, \quad \dots, \quad x_n = t^{\delta_n} + \dots$$

where

$$0 = \delta_1 < \delta_2 < \dots < \delta_n.$$

Proof: Starting with an arbitrary coordinate system we combine the following three elementary operations, which all amount to non-singular linear transformations:

1) Interchange of two coordinates: $x_i \to x_j$, $x_j \to x_i$ $(i<j)$.
2) Multiplication of a coordinate by a non-vanishing constant: $x_i \to cx_i$ $(c \neq 0)$.
3) Replacement of a coordinate x_j by the difference $x_j - bx_i$ where $i < j$ and b is an arbitrary constant.

(All coordinates except x_i, x_j in 1), x_i in 2), x_j in 3), are left unchanged.)

Since at least one of the power series $x_i(t)$ begins with a non-vanishing constant term, we can evidently combine the operations 1) and 2) so as to make $x_1(t) = 1+\dots$ By subtracting from $x_2(t),\dots,x_n(t)$ suitable multiples of $x_1(t)$ [operation 3)] we annihilate the constant terms of these n - 1 series. We may then write

$$x_2(t) = t^{V}\cdot x_2^*(t), \quad \dots, \qquad x_n(t) = t^{V}\cdot x_n^*(t)$$

with a positive exponent v while the constant term in at least one of the power series $x_2^*(t),\ldots,x_n^*(t)$ differs from zero. Assuming our theorem to be true for the reduced representation $(x_2^*(t),\ldots,x_n^*(t))$ in n - 1 dimensions, we obtain the desired result for the dimensionality n. We find n - 1 positive integers $v_1 = v, v_2,\ldots, v_{n-1}$ such that the exponents in our theorem equal

$$\delta_1 = 0, \qquad \delta_p = v_1 + \ldots + v_{p-1} \qquad (p=2,\ldots,n).$$

Hence

$$\delta_p - \delta_{p-1} = v_{p-1} \qquad (p=2,\ldots,n).$$

The geometric significance of the exponents δ may be described as follows. The plane (α), $\sum \alpha_i x_i = 0$, does not intersect the curve at t = 0 unless $\alpha_1 = 0$. Any plane satisfying this condition,

$$\alpha_2 x_2 + \ldots + \alpha_n x_n = 0,$$

intersects δ_2 times, and this is the exact multiplicity unless $\alpha_2 = 0$. If $\alpha_2 = 0$, then the plane

$$\alpha_3 x_3 + \ldots + \alpha_n x_n = 0$$

intersects δ_3 times and not more often unless $\alpha_3 = 0$; etc.

The non-homogeneous coordinates $x_2/x_1,\ldots,x_n/x_1$ all vanish at t = 0 to the order v_1. If one interprets the parameter t as <u>time</u>, the point t = 0 is a <u>stationary point</u> or <u>cusp</u> if and only if $v_1 > 1$, and for any value of v_1 it is reasonable to count this point as a (v_1-1)-fold cusp.

Let us now form the osculating p-ad $X^p = [x,x',\ldots, x^{(p-1)}]$. Since x_i begins with the terms t^{δ_i} the derivative $x_i^{(p)}$ begins with

$$\delta_i(\delta_i-1)\ \ldots\ (\delta_i-p+1)\ t^{\delta_i-p} = \phi_p(\delta_i)\ t^{\delta_i-p}$$

where $\phi_p(z)$ is the polynomial $z(z-1)\ \ldots\ (z-p+1)$. [If $\phi_p(\delta_i) = 0$, i. e. if $\delta_i < p$, it will actually begin with a later term.] Hence $X^p(i_1 \ldots i_p)$ begins with the term

$$(7.2)\quad \begin{vmatrix} \phi_0(\delta_{i_1}), & \ldots, & \phi_0(\delta_{i_p}) \\ \cdot & \cdot\ \cdot\ \cdot & \cdot \\ \phi_{p-1}(\delta_{i_1}), & \ldots, & \phi_{p-1}(\delta_{i_p}) \end{vmatrix} \cdot t^{d(i_1 \ldots i_p)} \quad \text{where}$$

$$d(i_1 \ldots i_p) = \delta_{i_1} + (\delta_{i_2}-1) + \ldots + (\delta_{i_p}-p+1),$$

and this is actually so because the determinant in front is the difference product of the distinct exponents $\delta_{i_1}, \ldots, \delta_{i_p}$ and hence $\neq 0$. Consequently all $X^p(i_1 \ldots i_p)$ contain the factor t^{d_p} where

$$d_p = d(1,2,\ldots,p) = \delta_1 + (\delta_2-1) + \ldots + (\delta_p-p+1),$$

and $X^p(1,\ldots,p)$ contains no higher power of t. After removing the common factor t^{d_p},

$$X^p(i_1 \ldots i_p) = t^{d_p} \cdot X^p_{red}(i_1 \ldots i_p),$$

we obtain the reduced representation $X^p_{red}(i_1 \ldots i_p)$ of $\mathfrak{C}_p$.

We have

$$d_1 = 0, \quad d_p - d_{p-1} = \delta_p - (p-1) \qquad (2 \leq p \leq n),$$

and after setting $X^o = 1$, $d_o = 0$, furthermore

$$(d_p - d_{p-1}) - (d_{p-1} - d_{p-2}) = (\delta_p - \delta_{p-1}) - 1 = v_{p-1} - 1$$

or

$$(7.3)\quad d_{p+1} - 2d_p + d_{p-1} = v_p - 1 \text{ for } p = 1,\ldots,n-1.$$

Thus the second differences of the d_p are non-negative.

We have determined the number of times, $\nu(\alpha)$, which the curve at $t = 0$ intersects the plane (α), $\sum \alpha_i x_i = 0$, by substituting a reduced representation $x_i = \mathfrak{P}_i(t)$ into the linear form $\sum \alpha_i x_i$. Should we use the non-reduced representation $x_i = t^d \mathfrak{P}_i(t)$ the order in which this form vanishes for $t = 0$ would turn out to be $d + \nu(\alpha)$. For higher ranks p we can execute the analogous steps by considering any linear form of the (ordered) components $X^p(i_1,\ldots,i_p)$ of the p-ad X^p,

$$(A^pX^p) = \sum_i{}' A(i_1\ldots i_p)X(i_1\ldots i_p) = \tag{7.4}$$

$$\begin{vmatrix} (\alpha_1 x) & ,\ \ldots\ , & (\alpha_p x) \\ (\alpha_1 x') & ,\ \ldots\ , & (\alpha_p x') \\ \cdot & \cdot\ \cdot\ \cdot & \cdot \\ (\alpha_1 x^{(p-1)}), & \ldots\ , & (\alpha_p x^{(p-1)}) \end{vmatrix}$$

whose coefficients constitute a special contravariant p-ad

$$A = A^p = [\alpha_1\ldots\alpha_p].$$

If the form does not vanish identically by the substitution $x_i = \mathfrak{P}_i(t)$, it will vanish in a certain order $d_p + \nu_p(A)$ for $t = 0$. It is essential to split off the common factor t^{d_p} of all $X^p(i_1,\ldots,i_p)$. The "intersection number" $\nu_p(A)$ is the order of vanishing at $t = 0$ of the form $(A^pX^p_{red})$ built with the reduced representation X^p_{red} of $\mathfrak{C}_p$. If we set $A^p = {*A^{n-p}}$ this number is the multiplicity with which the osculating p-element of the curve cuts a fixed (n-p)-element A^{n-p} at $t = 0$.

Let $p < n$. All ordered components $X^p(i_1\ldots i_p)$ with the exception of $X^p(1,\ldots,p)$ have $i_p \geq p+1$ and hence contain at least the power

$$d(1,\ldots,p-1,p+1) = d_p + (\delta_{p+1} - \delta_p) = d_p + \nu_p$$

of t as factor, and $X^p(1,\dots,p-1.p+1)$ contains exactly this power. Hence the common factor of all ordered components $X^p_{red}(i_1\dots i_p)$ with the exception of $X^p_{red}(1,\dots,p)$ is the power t^{v_p}. Therefore v_p plays the same role for the associated curve $\mathfrak{C}_p$ of the osculating p-elements as v_1 does for $\mathfrak{C} = \mathfrak{C}_1$. In particular, the inequality $v_p > 1$ indicates that the osculating p-element is stationary at t=0; we then say that t = 0 is a _stationary point of rank_ p, and whether $v_p > 1$ or $= 1$, we call $v_p - 1$ the pth _stationary index_. E.g. if $v_2 > 1$, then the _tangent_ is stationary at t = 0, and that point is an _inflection point_ of the curve. For higher p our language lacks equally suggestive terms.

In a certain neighborhood $0 < |t| < r$ of t = 0, the point t = 0 itself excluded, the determinant

$$X^p(1,\dots,p) \neq 0 \qquad (\text{for } p=0,\dots,n).$$

Hence no one of these points is stationary of rank p: _the stationary points of any rank are isolated_.

We turn now to the question of degeneracy. The above argument, if applied for p = n shows that the Wronskian of the non-degenerate curve is divisible by no higher power of t than by t^{d_n}. It is therefore not identically zero: the Wronskian cannot vanish identically unless the n functions $x_1(t), x_2(t), \dots, x_n(t)$ are linearly dependent. Thus we have established the converse of the evident proposition that the Wronskian of n analytic functions which are linearly dependent is identically zero. If exactly m of the n functions $x_1(t),\dots,x_n(t)$ are linearly independent, then $X^{m+1},\dots,X^n$ vanish while $X^1,\dots,X^m$ are different from zero. One sees this at once if one normalizes the coordinate system so that $x_{m+1}(t) = 0, \dots, x_n(t) = 0$; then $X^m(1,2,\dots,m)$ is the Wronskian of the m linearly independent functions $x_1(t), \dots, x_m(t)$. _Hence $\mathfrak{C}_{m+1}$ is totally degenerate if and only if $\mathfrak{C}$ is contained in a linear_ m-_subspace of_ $\mathfrak{S}$.

We proceed once more on our original assumption that $\mathfrak{C}$ is non-degenerate. We shall then show that a linear form (AX), (7.4), does not vanish identically in t unless $A = 0$. In this sense the associated curve $\mathfrak{C}_p$ is non-degenerate. Indeed, if $\alpha_1, \dots, \alpha_p$ are p linearly independent contravariant vectors, then

$$y_1 = (\alpha_1 x) , \dots , y_p = (\alpha_p x)$$

are p linearly independent analytic functions of t; hence their Wronskian, which is the determinant in the last member of (7.4), cannot vanish.

However, it may well happen that for a certain <u>non-special</u> A the linear form (AX) vanishes identically in t, even if $A \neq 0$. We may look upon the equation $X^p = X^p(t)$ as defining a curve $\mathfrak{C}_p$ in the projective $\binom{n}{p}$-space $\mathfrak{S}^{(p)}$ whose elements are the p-ads X^p. In this space the equation $(AX) = 0$ defines a plane if $A = A^p$ is any non-vanishing contravariant p-ad. The curve $\mathfrak{C}_p$ will lie in a certain linear subspace of $\mathfrak{S}^{(p)}$ of dimensionality N_p if N_p is the exact number of linearly independent functions among the ordered components $X^p(i_1 \dots i_p;t)$. From (7.2) it follows that N_p is at least as large as the number of numerically distinct p-sums

$$(7.5) \qquad \delta_{i_1} + \delta_{i_2} + \dots + \delta_{i_p} \qquad (i_1 < i_2 < \dots < i_p)$$

which may be derived from n given different numbers $\delta_1 < \delta_2 < \dots < \delta_n$. A fairly easy combinatorial argument shows that therefore

$$(7.6) \qquad N_p \geqq p(n-p) + 1.$$

Indeed, let us string p beads on a vertical wire with (equidistant) holding points 1,...,n running from top to bottom ($n \geq p$). The beads as they follow each other downwards, shall be labeled 1,...,p. They are movable and can occupy the nodes $i = 1,\dots,n$ on the wire. A "position" is described by the marks i_α which the several beads occupy, $i_1 < i_2 < \dots < i_p$. At the beginning of the game they shall be at the top: $i_1 = 1, \dots, i_p = p$. A simple "move" consists in moving one bead down one step while the others stay put. The problem is to bring, by a series of simple moves, all the beads to the bottom where they will occupy the last p marks. It can be done, for instance, by first dropping the lowest bead, then the next, and so on. The number of simple moves needed is obviously p(n-p) because each bead has finally moved down n - p steps. If $(i_1,\dots,i_p)$ are the p(n-p) + 1 consecutive positions encountered in any such game, the corresponding number (7.5) increases by each move.

Simple examples show that the lower limit (7.6) is actually obtained for some non-degenerate curves $\mathfrak{C}$. However, this entire question is of little interest to us. If we make it a rule never to admit any other than special p-ads, then a relation (AX) = 0 is impossible unless A = 0. Let us rejoice in this simple fact and turn our eyes away from the complications which arise if we let A roam beyond the confines of the special p-ads.

§8. Dual curve

Given a branch $x(t) = (x_1(t),\dots,x_n(t))$ of a non-degenerate analytic curve $\mathfrak{C}$, the dual curve $\mathfrak{C}^*$: $\xi(t) = (\xi_1(t),\dots,\xi_n(t))$ is defined by

$$(8.1) \qquad [x, x', \dots, x^{(n-2)}] = *\xi.$$

It is the locus of the osculating (n-1)-element if one represents this by its dual coordinates ξ. Remembering that the Wronskian W is given by

$$[x, x', \dots, x^{(n-1)}] = *W,$$

we have

$$(\xi x) = 0, \quad (\xi x') = 0, \quad \ldots, \quad (\xi x^{(n-2)}) = 0, \tag{8.2}$$
$$(\xi x^{(n-1)}) = W.$$

By differentiating the first equation,

$$(\xi' x) + (\xi x') = 0,$$

and taking the second into account, one gets

$$(\xi' x) = 0;$$

in the same way, from the second and third,

$$(\xi' x') = 0,$$

etc:

$$(\xi' x) = 0, \ (\xi' x') = 0, \ \ldots, \ (\xi' x^{(n-3)}) = 0,$$
$$(\xi' x^{(n-2)}) = -W.$$

Differentiating again, and each time taking the following equation into account, one obtains similar relations for ξ'', and iteration of the process finally results in the following equations:

$$(\xi^{(i)} x^{(j)}) = \begin{cases} 0 & \text{if } i + j \leq n - 2, \\ (-1)^i W & \text{if } i + j = n - 1. \end{cases} \tag{8.3}$$

Multiplication of the two determinants

$$|\theta_0, \theta_1, \ldots, \theta_{n-1}| \cdot |x, x', \ldots, x^{(n-1)}|$$

gives the determinant of the products

$$(\theta_k x^{(h)}) \qquad (k,h = 0,1,\ldots,n-1).$$

We choose $\theta_0, \ldots, \theta_{n-p-1}$ as indeterminates, but put

$$\theta_{n-p} = \xi, \ \theta_{n-p+1} = \xi', \ \ldots, \ \theta_{n-1} = \xi^{(p-1)}.$$

Then by the relations (8.3) the last-mentioned determinant is factored into

$$\det(\theta_r x^{(s)})\cdot\det(\xi^{(i)}x^{(n-p+j)}) \qquad \begin{bmatrix} r,s=0,\ldots,n-p-1 \\ i,j=0,\ldots,p-1 \end{bmatrix},$$

the second factor having the value

$$\begin{vmatrix} 0, & 0, & \ldots, & 0, & W \\ 0, & 0, & \ldots, & -W, & * \\ \cdot & \cdot & \cdot & \cdot & \cdot \\ (-1)^{p-1}W, & *, & \ldots, & *, & * \end{vmatrix} = W^p.$$

After cancelling the factor $W \neq 0$ one finds

$$[\theta_0,\ldots,\theta_{n-p-1}, \xi,\ldots,\xi^{(p-1)}] = W^{p-1}\cdot\det(\theta_r x^{(s)}).$$

Hence according to the definition of the *-operation

$$(8.4)\quad *[\xi,\ldots,\xi^{(p-1)}] = W^{p-1}\cdot[x,x',\ldots,x^{(n-p-1)}].$$

This equation at which we have been aiming shows that not only $\mathfrak{C}_{n-1}$ and $\mathfrak{C}^* = \mathfrak{C}_1^*$, but also $\mathfrak{C}_{n-p}$ and $\mathfrak{C}_p^*$ are related by the *-operation.

Our argument supposes non-degeneracy of the original curve, i.e. linear independence of $x_1(t), \ldots, x_n(t)$, or $W(t) \neq 0$. The highest case $p = n$ of our formula shows that then the Wronskian of $\xi_1(t), \ldots, \xi_n(t)$ is likewise different from zero.

Although we are going to use the relation (8.4) for non-degenerate curves $\mathfrak{C}$ only, our algebraic conscience prompts us to state the fact that it holds whether $W \neq 0$ or $W = 0$. By successive differentiation of

$$\xi_n = \begin{vmatrix} x_1, & \ldots, & x_{n-1} \\ x_1', & \ldots, & x_{n-1}' \\ \cdot & \cdot & \cdot \\ x_1^{(n-2)}, & \ldots, & x_{n-1}^{(n-2)} \end{vmatrix}.$$

one finds expressions for $\xi_n, \xi'_n, \ldots, \xi_n^{(p-1)}$ in terms of $x_i, x'_i, \ldots, x_i^{(n+p-3)}$. The same for the other components $\xi_i, \xi'_i, \ldots, \xi_i^{(p-1)}$. If we define $\xi, \xi', \ldots, \xi^{(p-1)}$ by these expressions, then the equations (8.3) up to $i = p - 1$ are algebraic identities in the independent variables $x_i, x'_i, \ldots, x_i^{(n+p-3)}$, and in the same sense we obtain by our proof the algebraic identity

$$(8.5)\quad W\cdot{*}[\xi,\xi', \ldots, \xi^{(p-1)}] = W^p\cdot[x,x', \ldots, x^{(n-p-1)}].$$

But W is a function of the independent variables $x_i, x'_i, \ldots, x_i^{(n-1)}$ which does not vanish identically, and hence by a general principle of algebra one may cancel the factor W on both sides of (8.5). Thus our algebraic interpretation frees us from the assumption $W(t) \neq 0$.

If $\mathfrak{C}$ is given by a reduced representation, the corresponding representation of $\mathfrak{C}^*$ will be of the form $\xi_i = t^{d_{n-1}}\cdot\Pi_i(t)$ where the representation by the n power series $\Pi_i(t)$ is reduced. Let d^*, v^*, ν^* denote the quantities which have the same significance for $\mathfrak{C}^*$ as d, v, ν have for $\mathfrak{C}$. Then the relation (8.4) and the above remark show that

$$d_p^* = (p-1)d_n - pd_{n-1} + d_{n-p}.$$

Forming first and second differences, we find

$$\delta_p^* = \delta_n - \delta_{n-p+1},$$

$$(8.6)\qquad v_p^* = v_{n-p}.$$

Moreover the equation

$$({*}A,{*}\Xi) = (A\Xi) \text{ for } \Xi = [\xi, \ldots, \xi^{(p-1)}]$$

proves that

$$(8.7)\qquad \nu_p^*(A) = \nu_{n-p}(A)$$

where $A = A^p$ is any special covariant p-ad and $A = A^{n-p} = *A^p$. Thus we obtain the duality laws (8.7), (8.6) for order and stationarity index.

§9. Rational curves

From the local investigation of analytic curves we turn to their behavior in the large and as a first example consider the rational or "unicursal" curves. Such a curve is defined if the n coordinates $x_1, \ldots, x_n$ are set up as rational functions of a complex parameter z,

$$x_i = R_i(z) \qquad (i=1,\ldots,n).$$

Replacing $R_i(z)$ by $Q(z)\cdot R_i(z)$ does not change the curve, whatever the non-vanishing rational function $Q(z)$ which appears here as gauge factor. One could choose this gauge factor in such a way that the n rational functions $R_i(z)$ become polynomials without a common divisor. However convenient for certain purposes this normalization may be, it is not advisable to include it in the definition because it assigns an unduly exceptional role to the value $z = \infty$ of the parameter.

For any finite value $z = z_o$ we use $z - z_o = t$ as the local parameter. With a certain integral exponent d, which might be positive, zero, or negative, we shall have in the neighborhood of $z = z_o$:

$$(9.1) \quad R_1(z) = t^d \cdot \mathfrak{P}_1(t), \; \ldots \;, \quad R_n(z) = t^d \cdot \mathfrak{P}_n(t)$$

where the n power series $\mathfrak{P}_i(t)$ constitute a <u>reduced</u> local representation of our rational curve. For $z = \infty$ we choose $1/z = t$ as local parameter, and then obtain similar equations (9.1) for the neighborhood of $z = \infty$. Suppose the curve to be non-degenerate and a plane (α),

$$\sum \alpha_i x_i = 0 \qquad [(\alpha_1,\ldots,\alpha_n) \neq (0,\ldots,0)],$$

to be given. For each "spot" $\mathfrak{p}$, finite or infinite,

$z = z_0$ or $z = \infty$, the local reduced representation enables us to determine the multiplicity $\nu(\mathfrak{p};\alpha)$ of the intersection of the curve with the plane (α) at $\mathfrak{p}$; see the definition of $\nu(\alpha)$ in §7. The spots where this multiplicity is actually positive are isolated. Hence there is only a finite number of such spots on the z-sphere, and we can form the sum

$$\sum_{\mathfrak{p}} \nu(\mathfrak{p};\alpha) = \nu \tag{9.2}$$

extending over all spots $\mathfrak{p}$; ν counts how often the curve intersects with the plane (α). We maintain:

> A given rational curve intersects every plane the same number of times.

In other words, (9.2) is independent of α. The integer ν is called the <u>order</u> of the curve. Indeed, let $\sum \beta_i x_i = 0$ be another plane and consider the rational function

$$f(z) = \sum \alpha_i R_i(z) / \sum \beta_i R_i(z)$$

which is independent of the choice of the gauge factor $Q(z)$. The order of the rational function $f(z)$ at any spot $\mathfrak{p}$, $z = z_0$ or $z = \infty$, is clearly $\nu(\mathfrak{p};\alpha) - \nu(\mathfrak{p};\beta)$. One knows that the sum of the orders of a non-vanishing rational function at all spots $\mathfrak{p}$ equals zero. This remark yields the desired result,

$$\sum_{\mathfrak{p}} \{\nu(\mathfrak{p};\alpha) - \nu(\mathfrak{p};\beta)\} = 0 \text{ or } \sum_{\mathfrak{p}} \nu(\mathfrak{p};\alpha) = \sum_{\mathfrak{p}} \nu(\mathfrak{p};\beta).$$

The theorem is also true for degenerate curves under the provision that only such planes are taken into consideration as do not completely contain the curve. In this modified sense the straight line is a rational curve of order 1.

From $\mathfrak{C}$ we pass to $\mathfrak{C}_p$. As before we set

$$X^p = [\mathfrak{P}(t), \mathfrak{P}'(t), \ldots , \mathfrak{P}^{(p-1)}(t)]$$

where t is the local parameter for a given spot. The order for $t = 0$ of the quotient

$$(\mathsf{A}\, X^p)/(\mathsf{B}\, X^p) \tag{9.3}$$

formed by means of two special contravariant p-ads A and B is $\nu_p(\mathsf{A}) - \nu_p(\mathsf{B})$, or as we now write, putting the spot $\mathfrak{p}$ in evidence, $\nu_p(\mathfrak{p};\mathsf{A}) - \nu_p(\mathfrak{p};\mathsf{B})$. If we replace $\mathfrak{P}_i(t)$ by $R_i(z)$, the numerator and denominator of (9.3) are multiplied by t^{pd}; if we replace differentiation with respect to t by differentiation with respect to z, the numerator and denominator are multiplied by the same factor $(\frac{dt}{dz})^{1+\ldots+(p-1)}$. Hence our quotient does not change if we let

$$R_z^p = [R, \frac{dR}{dz}, \ldots , \frac{d^{p-1}R}{dz^{p-1}}]$$

take the place of X^p. But then it turns into a rational function of z, and hence its total order, i.e. the sum of its orders at all spots $\mathfrak{p}$, equals zero. Thus we find that

$$\sum_{\mathfrak{p}} \nu_p(\mathfrak{p};\mathsf{A}) = \nu_p$$

is independent of A. This number ν_p might well be called the order of the associated curve $\mathfrak{C}_p$, but we prefer to speak of it as the <u>order of rank</u> p of the original curve $\mathfrak{C}$. If $\mathfrak{C}$ is non-degenerate, the special p-ads A and B are restricted only by the condition that they do not vanish; in the degenerate case one has to require that neither (A, R_z^p) nor (B, R_z^p) vanish identically in z.

The orders and the numbers of stationary points of all ranks p are connected by formulas which for algebraic curves in 4-space were first discovered by Plücker. Let us choose a constant non-vanishing contravariant p-ad A^p

for every p = 0, 1, ... , n and then consider the quotient

$$(9.4)\qquad Z_t^p = \frac{(A^{p-1}X^{p-1})\cdot(A^{p+1}X^{p+1})}{(A^pX^p)^2} \qquad (p=1,\ldots,n-1).$$

Its order for t = 0 is

$$\{d_{p-1}+d_{p+1}-2d_p\} + \{\nu_{p-1}(A^{p-1}) + \nu_{p+1}(A^{p+1}) - 2\nu_p(A^p)\}.$$

The first summand equals the stationarity index $v_p - 1$. Of the two operations of replacing $\mathfrak{p}_1(t)$ by $R_1 = t^{d}\mathfrak{p}_1$ and d/dt by d/dz, the first has no effect on our quotient, whereas the second adds a factor dz/dt because

$$\{0+1+\ldots+(p-2)\} + \{0+\ldots+p\} - 2\{0+\ldots(p-1)\} = 1.$$

Hence the equation

$$\frac{(A^{p-1}R_z^{p-1})\cdot(A^{p+1}R_z^{p+1})}{(A^pR_z^p)^2}\cdot\frac{dz}{dt} = Z_t^p.$$

The first factor at the left is a rational function of z (which is not affected by the gauge factor Q(z)) and thus its total order equals zero. The second factor dz/dt equals 1 for the finite spots and $-1/t^2$ for the infinite spot (z = 1/t); hence its total order is -2. Therefore, by summation over all spots,

$$(9.5)\qquad \sigma_p + (\nu_{p+1}-2\nu_p+\nu_{p-1}) = -2 \qquad (p=1,\ldots,n-1)$$

where

$$\sigma_p = \sum_{\mathfrak{p}} v_p(\mathfrak{p})-1\}$$

is the total number of stationary points of rank p. <u>These are the generalized Plücker formulas for unicursal curves.</u>

The dual curve of a rational curve is again rational.

Distinguishing its characteristic numbers by an asterisk, we have

$$\sigma^*_p = \sigma_{n-p}, \quad \nu^*_p = \nu_{n-p}.$$

Up to now we have assumed a rational function to be defined by an explicit formal expression, namely as the quotient of two polynomials. We have outlined in the Introduction how this formal expression leads to the fundamental law of vanishing total order for any rational function which is not identically zero. However, Riemann taught us a less formal interpretation of rational functions: they are the functions which are regular analytic on the entire z-sphere (including the point $z = \infty$), except at isolated points where they have poles. We may define the rational curves in a similar fashion: A rational curve $\mathfrak{C}$ is given if with every spot $\mathfrak{p}$ on the z-sphere there is associated a branch of an analytic curve

$$(9.6) \qquad x_1 = \mathfrak{P}_1(t), \ldots, \quad x_n = \mathfrak{P}_n(t)$$

expressed in terms of the local parameter, $t = z - z_0$ or $1/z$, respectively, by means of a reduced representation; it is further required that the branches associated with the spots in a certain neighborhood of $\mathfrak{p}$ arise from the branch at $\mathfrak{p}$ by immediate analytic continuation. As before it is considered an integral part of the definition of the branch that it stays unaltered under multiplication of the n power series $\mathfrak{P}_i(t)$ by a common factor $\rho(t)$, this gauge factor being an arbitrary power series with non-vanishing constant term

$$\rho(t) = \rho_0 + \rho_1 t + \ldots \qquad (\rho_0 \neq 0)$$

which converges in a circle of positive radius around the origin.

Under these circumstances a linear form $\alpha_1 x_1 + \ldots + \alpha_n x_n$ does not determine an analytic function on the z-sphere by the substitution $x_1 = \mathfrak{P}_1(t)$, because of the arbitrary gauge factor $\rho(t)$. But the gauge factor cancels out in the quotient of two linear forms, and hence this quotient is a function on the z-sphere which is regular except for poles and therefore a rational function of z.

Viewed in this way our theory of rational curves at once suggests its generalization to arbitrary Riemann surfaces.

§10. Analytic and algebraic curves

A Riemann surface $\mathfrak{R}$ is a (connected) topological space in which the notion of analytic function is defined in such a manner as to enjoy locally the characteristic properties of an analytic function of a complex variable. More precisely, for any point $\mathfrak{p}_0$ of $\mathfrak{R}$ there is supposed to exist among the functions which are analytic at $\mathfrak{p}_0$ one, $t(\mathfrak{p})$, called a local parameter for $\mathfrak{p}_0$, with the following properties. It maps a certain neighborhood $\mathfrak{N}(\mathfrak{p}_0)$ of $\mathfrak{p}_0$ topologically onto the interior of a circle $|t| < a$ of positive radius in the complex t-plane. $t - t_1 = t - t(\mathfrak{p}_1)$ is a local parameter for any point $\mathfrak{p}_1$ in this neighborhood. Any power series in t,

$$f = c_0 + c_1 t + c_2 t^2 + \dots , \tag{10.1}$$

which converges in a circle $|t| < a'$ of positive radius $a' \leq a$ defines a function which is analytic at $\mathfrak{p}_0$; and vice versa, any function $f(\mathfrak{p})$ defined in a neighborhood of $\mathfrak{p}_0$ and analytic at $\mathfrak{p}_0$ is expressible in terms of the local parameter by such a power series (10.1), the equation holding good in a certain neighborhood $\subset \mathfrak{N}(\mathfrak{p}_0)$ of $\mathfrak{p}_0$.

According to this description, any local parameter s for $\mathfrak{p}_0$ must be of the form

$$s = b_1' t + b_2' t^2 + \dots \tag{10.2}$$

while, vice versa, t is expressible in terms of s in similar manner,

(10.2') $$t = b_1 s + b_2 s^2 + \dots$$

These two equations imply $b_1 b_1' = 1$; hence $b_1' \neq 0$ ($b_1 \neq 0$). One knows that (10.2) being an arbitrary power series with $b_1' \neq 0$, the analytic function $s(t)$ has a single-valued inverse $t(s)$ in a certain neighborhood of the origin expressible in the form (10.2'). From this it follows that any s which arises from t by a power series (10.2) with $b_1' \neq 0$ converging in a circle of positive radius might serve as local parameter. This type of parameter transformations has been anticipated in §6. Relations established in terms of a local parameter are independent of its choice if they are invariant with respect to all transformations (10.2), (10.2').

An evident implication of our description is the fact that a function analytic at $\mathfrak{p}_0$ is also analytic at all points $\mathfrak{p}_1$ in a certain neighborhood of $\mathfrak{p}_0$.

It is reasonable to say that t maps the neighborhood $\mathfrak{N}(\mathfrak{p}_0)$ mentioned above conformally upon the circle $|t| < a$ in the t-plane, and we therefore speak of $\mathfrak{N}(\mathfrak{p}_0)$ as a conformal neighborhood of $\mathfrak{p}_0$ (or conformal circle around $\mathfrak{p}_0$). The closed set which is mapped by t into a circle $|t| \leq a'$ of smaller radius $a' < a$ is said to be a conformal disk.

The general notion of a Riemann surface would be of little use unless we impose a certain condition of topological nature which may vaguely be described as exhaustibility. One who is concerned about making the Riemann surfaces accessible to the methods of combinatorial topology will formulate it more precisely as the possibility of triangulation. But for our purposes, in particular for the theory of integrals on a Riemann surface, covering of the surface by conformal disks rather than dissection into triangles will be found sufficient. Incidentally, T. Radó has shown that for Riemann surfaces the seemingly weaker condition of coverage by disks implies "triangulability". The central notion in this discussion is that of compactness.

A point set G in a topological space is said to be compact if it is known that whenever to each point $\mathfrak{p}$ of G there is assigned a set $U(\mathfrak{p})$ of which $\mathfrak{p}$ itself is an interior point, then one can select a finite number of points $\mathfrak{p}_1, \ldots, \mathfrak{p}_h$ in G such that the corresponding sets $U(\mathfrak{p}_1), \ldots, U(\mathfrak{p}_h)$ cover G. It is a fact (Heine-Borel) that a circular disk $|t| \leq a$ in the t-plane is compact. --A closed subset G_1 of a compact set G is compact. Indeed, assume that for each point $\mathfrak{p}$ of G_1 we are given a set $U(\mathfrak{p})$ of which $\mathfrak{p}$ is an interior point. For each point $\mathfrak{p}$ in the complement $\overline{G}_1$ of G_1 we determine a neighborhood $U(\mathfrak{p})$ of $\mathfrak{p}$ which does not penetrate into G_1; this is possible because G_1 is closed and therefore $\overline{G}_1$ open. Operating with this assignment $U(\mathfrak{p})$ for points $\mathfrak{p}$ inside and outside G_1, and making use of the compactness of G we prove that a finite number among the $U(\mathfrak{p})$ assigned to points $\mathfrak{p}$ of G_1 can be selected so as to cover G_1. -- A sub-set G_1 of a set G is said to be isolated in G if to every point $\mathfrak{p}$ of G one can assign a neighborhood $\mathfrak{N}(\mathfrak{p})$ which contains no point of G_1 with the possible exception of the center $\mathfrak{p}$ itself. An isolated subset of a compact set contains no more than a finite number of points.* Indeed, any selection of the neighborhoods $\mathfrak{N}(\mathfrak{p})$ just introduced which covers G must of necessity contain the $\mathfrak{N}(\mathfrak{p}_1)$ of each $\mathfrak{p}_1 \in G_1$.

After these preliminaries we formulate the general restriction imposed upon Riemann surfaces $\mathfrak{R}$. We assume that $\mathfrak{R}$ can be covered by a denumerable sequence of conformal disks $K_1, K_2, \ldots$. If this is so, and if to

* There has been a shift of terminology in recent years. Five years ago "compact" denoted this property of having no isolated subsets consisting of more than a finite number of points, whereas the word "bicompact" was in use for what we now call "compact".

every point $\mathfrak{p}$ of $\mathfrak{R}$ there is assigned a set $U(\mathfrak{p})$ of which $\mathfrak{p}$ is an interior point, then one can select a denumerable sequence of points $\mathfrak{p}_1, \mathfrak{p}_2, \ldots$ such that the corresponding sets $U(\mathfrak{p}_i)$ $(i=1,2,\ldots)$ cover the Riemann surface. In particular, this is the case when each $U(\mathfrak{p})$ is a circular disk of center $\mathfrak{p}$. In order to prove our statement one simply has to make use of the fact that each of the covering disks K_i is compact. If $\mathfrak{R}$ can be covered by a finite number of conformal disks $K_1, \ldots, K_h$, then it follows by the same argument that $\mathfrak{R}$ is compact, and in particular that, a disk $U(\mathfrak{p})$ of center $\mathfrak{p}$ being assigned to each point $\mathfrak{p}$, a finite number of these $U(\mathfrak{p})$ will suffice to cover $\mathfrak{R}$. Thus we arrive at the fundamental distinction of compact and non-compact Riemann surfaces.

Let $\mathfrak{R}$ be covered by the sequence $K_1, K_2, \ldots$ of conformal disks, and form the join $G_q = K_1 \cup K_2 \cup \ldots \cup K_q$ of the first q of them; G_q is compact, bounded by piecewise analytic contours, increases with q, and will exhaust $\mathfrak{R}$ with q tending to infinity. If $\mathfrak{R}$ is compact, G_q will actually coincide with the whole $\mathfrak{R}$ from a certain q on, whereas in case of non-compactness $\mathfrak{R}$ will not be covered completely at any finite stage.*

After this topological digression we return to the discussion of analytic functions on a given Riemann surface $\mathfrak{R}$. Being given a Riemann surface $\mathfrak{R}$, we know what it means that a function defined in a neighborhood of a point $\mathfrak{p}_0$ of $\mathfrak{R}$ is regular analytic at $\mathfrak{p}_0$. It is also clear under what condition a function f defined in a

---

* Concerning the definition of Riemann surfaces, see:

H. Weyl, Die Idee der Riemannschen Fläche, Leipzig, 1913 (2. Aufl. 1923).

P. Koebe, Abhandlungen zur Theorie der konformen Abbildung III, Jour. r. u. ang. Math. 147, 1917, 67-104, in particular pp. 70-73.

T. Radó, Ueber den Begriff der Riemannschen Fläche, Act. Litt. ac Sc. Szeged, 2, 1925, 101-121.

S. Stoilow, Leçons sur les principes topologiques de la théorie des fonctions analytiques, Paris, 1938.

neighborhood of $\mathfrak{p}_0$ except at $\mathfrak{p}_0$ itself has a pole at $\mathfrak{p}_0$; namely, if it is expressible in terms of a local parameter t in the form

$$f = a_{-h}t^{-h} + \dots + a_{-1}t^{-1} + a_0 + a_1 t + \dots .$$

The pole is of order h and the function has the order -h at $\mathfrak{p}_0$ if $a_{-h} \neq 0$. These notions are independent of the choice of the local parameter. A function which is regular or has at most a pole at $\mathfrak{p}_0$ is called <u>meromorphic at</u> $\mathfrak{p}_0$. Such a function f if not vanishing identically possesses an expansion

$$(10.3) \qquad f = a_h t^h + a_{h+1} t^{h+1} + \dots \qquad (a_h \neq 0)$$

with a first non-vanishing coefficient a_h. The integral exponent h of the initial term, which may be positive, zero, or negative, is said to be the <u>order</u> of f at $\mathfrak{p}_0$. A function f on the Riemann surface is called <u>meromorphic</u> if it is meromorphic everywhere.

In case the Riemann surface is compact, the number of poles must then be finite because the poles are of necessity isolated points. More completely, there is only a finite number of points $\mathfrak{p}$ where the order of the meromorphic function differs from zero. <u>The sum of the orders at all points $\mathfrak{p}$ equals zero</u>. This fundamental theorem, a particular case of the general residue theorem, follows if one dissects $\mathfrak{R}$ by the peripheries of a finite covering set of conformal disks and expresses the sum of orders of f in each piece as the integral along its contour of the logarithmic derivative df/f.

Once in possession of a non-constant meromorphic function $z(\mathfrak{p})$ on a given Riemann surface we may transform that surface, whether it is compact or not, into a covering surface over the complex z-plane (including the point $z = \infty$) by agreeing that a point $\mathfrak{p}$ lies over the point z_0 of the z-plane if $z(\mathfrak{p}) = z_0$. In case the surface $\mathfrak{R}$ is compact, the function $z(\mathfrak{p}) - z_0$ takes on the value 0 as often as ∞, whatever the given value z_0; i.e. the function $z(\mathfrak{p})$ takes on every value including ∞ the same number of times, let us say N times. Therefore $\mathfrak{R}$ is turned into a covering surface of N sheets over the z-plane. The meromorphic functions on $\mathfrak{R}$ then form a field of algebraic functions of the variable z; the field itself is of degree N.

Let $\mathfrak{R}$ be a given Riemann surface. An analytic curve of type $\mathfrak{R}$ in projective n-space is defined by associating with every point $\mathfrak{p}_0$ of $\mathfrak{R}$ a branch

$$(10.4) \qquad x_i = \mathfrak{P}_i(t) = c_i + c_i' t + \dots ,$$
$$(c_1, \dots , c_n) \neq (0, \dots , 0).$$

For this reduced representation, a local parameter t at $\mathfrak{p}_0$ is chosen, but it is understood that the choice of the local parameter is arbitrary and that the branch does not change if one multiplies the power series $\mathfrak{P}_i(t)$ by a common gauge factor $\rho(t)$ of the nature described before. Finally there is supposed to exist a neighborhood of $\mathfrak{p}_0$, $|t| < r$, such that the branches associated with points $\mathfrak{p}$ in this neighborhood arise from the branch (10.4) at $\mathfrak{p}_0$ by immediate analytic continuation. To be quite explicit, if the branch at $\mathfrak{p}$ is given in a definite representation, its n representing power series are required to be proportional to the ones obtained from (10.4) by immediate analytic continuation.

Let α_i be constants such that the linear form $\sum \alpha_i x_i$ vanishes by the substitution (10.4) identically in t (critical form). This form will then be annulled not only by the branch at $\mathfrak{p}_0$ but by the branch associated with any other point. This follows at once by analytic continuation because the Riemann surface is connected or, as Weierstrass would say, because we are concerned with a monogenic analytic curve $\mathfrak{C}$. Geometrically we should express the vanishing of our linear form by saying that the curve $\mathfrak{C}$ lies in the plane $\sum \alpha_i x_i = 0$. More generally, the curve $\mathfrak{C}$ lies entirely within an m-subspace if one of its branches does. We assume that $\mathfrak{C}$ is not totally degenerate, i.e. that not every linear form vanishes "along" $\mathfrak{C}$. The curve is non-degenerate if $\sum \alpha_i x_i$ vanishes along $\mathfrak{C}$ only if all coefficients α_i vanish.

The quotient of two non-critical linear forms

$$\sum \alpha_i x_i / \sum \beta_i x_i \tag{10.5}$$

is clearly a non-vanishing meromorphic function on $\mathfrak{R}$. In particular, if the coordinate x_n does not vanish identically on $\mathfrak{C}$, then the non-homogeneous coordinates

$$x_1/x_n,\ x_2/x_n,\ \dots\ ,\ x_{n-1}/x_n$$

are meromorphic functions on $\mathfrak{R}$. One could have defined the concept of an analytic curve of type $\mathfrak{R}$ in this manner. However, we have preferred to introduce this notion independently of, and on an equal footing with, that of an analytic function; from our standpoint it is more natural to consider a meromorphic function as the special case $n = 2$ of an analytic curve.

Following the same procedure as for rational curves, we study linear forms $(A^p X^p)$ of the osculating p-element $X^p = [x, x', \dots , x^{(p-1)}]$ and form by (9.4) the quotient Z_t^p (for $p=1,\dots,n-1$). The latter is not affected by the gauge factor $\rho(t)$, and under transformation of the local parameter, $t = t(s)$, transforms according to the equation

$$Z_s = Z_t \cdot \frac{dt}{ds}. \tag{10.6}$$

Z is thus a differential rather than a function on $\mathfrak{R}$. Let us explain this in detail!

A <u>function</u> z on $\mathfrak{R}$ assigns to an arbitrary point $\mathfrak{p}_0$ a value $z(\mathfrak{p}_0)$ which is independent of the local parameter. If $z(\mathfrak{p})$ is meromorphic it has an expansion (10.3) in the neighborhood of $\mathfrak{p}_0$. A <u>differential</u> dZ on $\mathfrak{R}$ assigns to an arbitrary $\mathfrak{p}_o$ a value $Z_t(\mathfrak{p}_o)$ which depends on the local parameter t in such a fashion that with respect to any two local parameters t and s the equation

$$Z_s(\mathfrak{p}_o) = Z_t(\mathfrak{p}_o)\cdot(\frac{dt}{ds})_o$$

holds. [In the notation (10.2'), $(\frac{dt}{ds})_o$ is the first coefficient b_1.] We therefore write

$$dZ = Z_t \cdot dt = Z_s \cdot ds.$$

For every point $\mathfrak{p}_1$ in a circular neighborhood $|t| < r$ of $\mathfrak{p}_o$ we may use $t - t_1 = t - t(\mathfrak{p}_1)$ as local parameter and thus form

$$Z_{t-t_1}(\mathfrak{p}_1) = \frac{dZ}{dt}(\mathfrak{p}_1).$$

The differential dZ is meromorphic at $\mathfrak{p}_o$ if $\frac{dZ}{dt}(\mathfrak{p}_1)$ permits an expansion like

$$a_h t_1^h + a_{h+1} t_1^{h+1} + \dots , \qquad \text{h integral,}$$

so that we may write in the neighborhood of $\mathfrak{p}_o$:

$$dZ = (a_h t^h + a_{h+1} t^{h+1} + \dots) \, dt.$$

Its order is h provided $a_h \neq 0$. dZ is meromorphic if it is of this type everywhere. If z is a meromorphic function, then dz is clearly a meromorphic differential. Vice versa, one obtains a function on $\mathfrak{R}$ by integrating a meromorphic differential, $\int dZ$, along arbitrary paths. But in general this "Abelian integral" will not be single-valued on $\mathfrak{R}$. We are here concerned with the differentials and not with their integrals. The quotient of any two meromorphic differentials, of which the second does not vanish identically, is a meromorphic function. After introducing this fundamental notion of differential we can indeed claim the expression (9.4) as a meromorphic differential on $\mathfrak{R}$.

Let us now study the case of a compact $\mathfrak{R}$. Analytic curves stemming from a compact $\mathfrak{R}$ are said to be algebraic. As previously mentioned, the total order of a meromorphic function on such an $\mathfrak{R}$ is zero. Applying this to the meromorphic function (10.5) we find that the total number of intersections of $\mathfrak{C}$ with a plane (α),

$$\sum_{p} \nu(p;\alpha) = \nu$$

is independent of (α). This number ν is again called the order of the algebraic curve.

A similar consideration applies to the quotient (9.3), which is not only independent of the gauge factor but also of the choice of the local parameter and hence a meromorphic function. We thus find

$$\sum_{p} \nu_{p}(p;A) = \nu_{p} \tag{10.7}$$

to be independent of A and obtain the order ν_p of rank p of our algebraic curve.

So far the argument is exactly the same as for rational curves. In order to obtain the Plücker formulas a slight change is required due to the lack of an analogue of the meromorphic function z on the z-sphere. Since the quotient of two meromorphic differentials is a meromorphic function, the difference of their total orders is zero; which means that every meromorphic differential on $\mathfrak{R}$ has the same total order. We denote that order by $2g - 2$. In the literature on Riemann surfaces one finds proofs for the fact that the "genus" g is a non-negative integer, and that 2g is the Betti number of the topological manifold $\mathfrak{R}$. If $\mathfrak{R}$ is, in particular, the z-sphere, then $g = 0$. Indeed dz has no zeros but a pole of order 2 at infinity.

The total order of the differential (9.4) equals

$$\sigma_p + (\nu_{p+1} - 2\nu_p + \nu_{p-1}).$$

Hence we obtain the following Plücker formulas for algebraic curves

$$\sigma_p + (\nu_{p+1} - 2\nu_p + \nu_{p-1}) = 2g - 2 \qquad [p=1,\ldots,n-1]. \tag{10.8}$$

The fact that (10.7) is independent of A and these Plücker formulas will be referred to as the first and the second main theorems for algebraic curves; it will be one of our main tasks to carry them over to arbitrary analytic curves.

§11. Meromorphic curves

Just as the z-sphere with inclusion of the point $z = \infty$ is the simplest example of a compact Riemann surface, so is the z-plane without the point $z = \infty$, which consists of all complex numbers z, the simplest non-compact example. When one speaks of meromorphic functions without mentioning any particular Riemann surface, one thinks of meromorphic functions in this "open" z-plane, and the corresponding analytic curves go under the name of meromorphic curves. The common reason for the simplicity of this example, as well as that of the rational curves, is that a single complex coordinate serves to identify the points of the entire surface; quite unlike the general Riemann surface which is a patchwork of pieces each of which carries its separate coordinate.

We may therefore define the meromorphic curves, just as the rational curves, in two different ways, either globally or by analytic continuation from point to point. The first more elementary and formal definition consists in setting up the n homogeneous coordinates x_i as meromorphic functions of z,

$$x_1 = f_1(z), \ \ldots\ , x_n = f_n(z), \tag{11.1}$$

with the understanding that the relations

$$x_1 = q(z)\cdot f_1(z), \ \ldots\ , x_n = q(z)\cdot f_n(z)$$

define the same curve whatever the non-vanishing meromorphic function $q(z)$. We exclude the case that all $f_i(z)$ vanish identically. If, in particular, $f_n(z) \neq 0$ one might replace our definition by setting up the non-homogeneous coordinates $x_1/x_n, \ldots, x_{n-1}/x_n$ as meromorphic functions, from which the arbitrary factor of proportionality $q(z)$ has disappeared. In this sense a single meromorphic function, written in the form x_1/x_2, is equivalent to a meromorphic curve in 2-space ($n = 2$).

One may use the factor $q(z)$ to normalize the representation of the curve. Consider an arbitrary circle $|z| \leqq R$ around the origin. In this circle the functions $f_1(z), \ldots, f_n(z)$ have but a finite number of poles $z_1, \ldots, z_N$. Hence we can determine a polynomial

$$Q(z) = (z-z_1)^{h_1} \ldots (z-z_N)^{h_N},$$

such that after multiplication of $f_1(z), \ldots, f_n(z)$ by $Q(z)$ these functions become <u>regular</u> throughout the circle $|z| \leqq R$. In similar manner we can remove common zeros which they may have in that circle. The result is a representation (11.1) for which the $f_i(z)$ are regular functions in the circle $|z| \leqq R$ without common zeros. Of course this representation is not uniquely determined. But any representation of this kind differs from the one we have constructed by a common meromorphic factor $q(z)$ which is regular and without zeros in $|z| \leqq R$.

One can go beyond this result and take hold of the whole plane instead of merely a circle of finite radius R by making use of Weierstrass's construction of entire functions (see Introduction). The poles of the functions $f_1(z), \ldots, f_n(z)$ are isolated points and hence may be ordered according to increasing moduli,

$$|z_1| \leqq |z_2| \leqq |z_3| \leqq \ldots, \quad \lim_{n \to \infty} |z_n| = \infty.$$

Let each pole occur in the sequence $z_1, z_2, \ldots$ as often as necessary; if for instance all the n products $(z-a)^5 \cdot f_i(z)$ are regular at $z = a$ then let a occur 5 times in that sequence. Weierstrass's construction yields an entire function $H(z)$ with the zeros $z_1, z_2, \ldots$. After multiplication of the n meromorphic functions $f_i(z)$ by this $H(z)$, the $f_i(z)$ themselves become entire functions. By a similar construction one may deprive them of all common zeros. The representation $x_i = f_i(z)$ by entire functions without common zeros thus obtained is determined but for a factor $q(z)$ which is an entire function without zeros and hence of the form $e^{\kappa(z)}$ where $\kappa(z)$ is entire. We do not propose to make use of this normalization in our systematic theory. But if we want to subsume the theory of entire functions $f(z)$ under that of meromorphic curves in 2-space, we naturally come upon a representation of this kind, namely:

$$x_1 = f(z), \quad x_2 = 1.$$

The less formal definition of a meromorphic curve by analytic continuation is nothing else but that special case of our general concept of analytic curve which corresponds to the open z-plane as Riemann surface. In this case it is not necessary to introduce a letter t besides z for the local parameter and we need not discuss any transformations of the local parameter, because for every point z_0 of our plane we simply choose $z - z_0$ as the local parameter. Hence a meromorphic curve is given if with every complex value z_0 there is associated a branch

$$x_i = \mathfrak{P}_i(z-z_0) = c_i + c_i'(z-z_0) + \ldots$$

represented by power series in $z - z_0$ whose initial terms do not all vanish,

$$(c_1, \ldots, c_n) \neq (0, \ldots, 0).$$

Again the stipulations of an arbitrary gauge factor

$$\rho(z-z_0) = \rho_0 + \rho_1(z-z_0) + \cdots \qquad (\rho_0 \neq 0)$$

and of immediate analytic continuation are to be added.

From this definition one falls back upon the first if one chooses n constants α_i^0 such that $x_0 = \sum \alpha_i^0 x_i$ is a non-critical linear form. Then

$$x_1/x_0 = f_1(z), \ \ldots \ , x_n/x_0 = f_n(z)$$

are meromorphic functions (which, by the way, satisfy the normalizing condition $\sum \alpha_i^0 f_i(z) = 1$), and (11.1) is a permissible representation of $\mathfrak{C}$ in the sense of our first definition.

Rational curves are special meromorphic curves, but in studying them as such we have to exclude the point $z = \infty$.

CHAPTER II

FIRST MAIN THEOREM FOR MEROMORPHIC CURVES

§1. The condenser formula

We exhaust the z-plane by the circles K_r, $|z| \leq r$, around the origin, letting the radius r increase from 0 to ∞. Whereas the total order of a rational function of z is zero if the point $z = \infty$ is included, we can express the total order n(r) of a given meromorphic function f(z) within the circle K_r by a contour integral along the circumference k_r of K_r. The negative of it will tend with $r \to \infty$ to the order of f(z) at $z = \infty$ in case f is rational; hence in the general meromorphic case it plays the role of a compensating term which for $r \to \infty$, as it were, makes up for the non-existing point $z = \infty$ whose influence is spread over the whole circumference of the circle of large radius r.

Let us denote by $\nu(a)$ the order of f(z) at the arbitrary point z = a and assume that no zero or pole lies on the circumference k_r. Then the total order

$$n(r) = \sum_{|z|<r} \nu(z)$$

in the circle K_r equals $1/2\pi i$ times the integral along k_r of the logarithmic derivative of f,

$$(1.1) \qquad \frac{1}{2\pi i}\int_{k_r} df/f = \frac{1}{2\pi}\int_{\vartheta=0}^{2\pi} d\{\Im \log f(re^{i\vartheta})\}.$$

A line element tangential to k_r measures $r d\vartheta$, one in normal direction dr; hence by the Cauchy-Riemann differential equations

$$\frac{1}{r}\cdot\frac{\partial}{\partial\vartheta}(\Im \log f) = \frac{\partial}{\partial r}(\Re \log f) = \frac{\partial}{\partial r}\log|f|,$$

and (1.1) changes into

$$(1.2) \qquad \frac{n(r)}{r} = \frac{d}{dr} \{\frac{1}{2\pi} \int_0^{2\pi} \log |f(re^{i\vartheta})| \cdot d\vartheta\}.$$

The fact that the right member is a derivative with respect to r suggests integration of this formula. Because of the denominator r at the left, it will not do to extend the integration down to r = 0 unless one is sure that f has neither a zero nor a pole at the origin, and it is inadvisable to exclude this possibility. We therefore integrate from a positive value $r = r_0$ fixed once for all, up to an arbitrary (variable) radius $R > r_0$. Setting

$$(1.3) \qquad \int_{r_0}^{R} n(r) \cdot \frac{dr}{r} = N(R)$$

we obtain a formula fundamental for all our subsequent investigations, viz.

$$(1.4) \qquad N(R) - [\frac{1}{2\pi} \int_0^{2\pi} \log |f(re^{i\vartheta})| d\vartheta]_{r=r_0}^{R} = 0.$$

Haven't we jumped somewhat hastily to this conclusion? The equation (1.2) holds for r only if there lie no zeros and poles of f on the circle k_r. When in blowing up the circle k_r one passes a zero or pole of f the function n(r) jumps. This singularity is harmless as far as the integration (1.3) is concerned, but formula (1.4) will be correct only if we can show that the mean value

$$\mathfrak{M}_r \log |f| = \frac{1}{2\pi} \int_0^{2\pi} \log |f(re^{i\vartheta})| \cdot d\vartheta$$

is a continuous function of r, even for those critical values of r. This point ought to be investigated scrupulously.

We split off the zeros a and poles b of the function f(z) in the circle $|z| \leq R$ and thus write

$$f(z) = \{\prod(z-a)/\prod(z-b)\} \cdot g(z).$$

There is no difficulty in applying the formula (1.2) to g for all values $r \leqq R$, with the result that

$$\mathfrak{M}_r \log |g| = \text{const.} \qquad \text{for } r \leqq R.$$

Because

$$\log |f| = \{\sum_a \log |z-a| - \sum_b \log |z-b|\} + \log |g|,$$

it is therefore enough to establish the formula (1.4) for the particular function $f(z) = z - a$. We are going to prove:

$$(1.5) \qquad \mathfrak{M}_r \log |z-a| = \psi_a(r)$$

where $\psi_a(r) = \log |a|$ for $r \leqq |a|$ and $\psi_a(r) = \log r$ for $r \geqq |a|$. $\psi_a(r)$ is clearly a continuous function of r even for r = a.

Indeed, if $|z| < |a|$, we write $u = z/a$ and

$$\log |z-a| = \log |a| + \Re \log (1-u).$$

But the integral

$$(1.6) \qquad \int \log (1-u) \cdot \frac{du}{u}$$

extended over a circle $|u| = r < 1$ vanishes, as follows at once from the expansion

$$- \frac{\log (1-u)}{u} = \frac{1}{1} + \frac{u}{2} + \frac{u^2}{3} + \dots .$$

If $|z| > |a|$ we write

$$\log |z-a| = \log |z| + \Re \log (1-u)$$

where this time $u = a/z$, and then use the same evaluation of (1.6).

This suffices for proving our formula (1.4) at least if r_0 and R are no critical values. As it should be, the continuous function $\psi_a(r)$ has the derivative $n_a(r)/r$ with $n_a(r) = 0$ for $r < |a|$, $n_a(r) = 1$ for $r > |a|$. The limiting case $r = a$ of (1.5) is a different matter because then we are dealing with an improper integral. If we wish to include it we have to show that the improper integral

$$\int_0^{2\pi} \log |1-e^{i\vartheta}| \cdot d\vartheta$$

equals zero. The function $\frac{\log (1-z)}{z}$ is regular in the z-plane slit along the line $\Im z = 0$, $\Re z \geqq 1$. We describe a circle κ of small radius δ around 1, $\delta = 2 \sin \frac{1}{2}\delta'$. The part $\delta' \leqq \vartheta \leqq 2\pi - \delta'$ of the unit circle outside κ can be made into a circuit by adding the inner arc κ' of κ. Apply Cauchy's integral theorem to the above function and this circuit. On κ'

$$|\log (1-z)| \leq \sqrt{\{(\log \delta)^2 + (\tfrac{\pi}{2})^2\}}, \qquad |z| \geqq 1-\delta ,$$

and thus

$$\left|\frac{1}{2\pi} \int_{\kappa'} \log (1-z) \cdot \frac{dz}{z}\right| \leqq \frac{1}{2} \frac{\delta}{1-\delta} \sqrt{\{(\log \delta)^2 + (\tfrac{\pi}{2})^2\}}.$$

Hence we obtain the same upper bound for the absolute value of $\frac{1}{2\pi} \int_{\delta'}^{2\pi-\delta'} \log (1-e^{i\vartheta}) \cdot d\vartheta$ and <u>a fortiori</u> for the absolute value of its real part

$$\frac{1}{2\pi} \int_{\delta'}^{2\pi-\delta'} \log |1-e^{i\vartheta}| \cdot d\vartheta.$$

This upper bound tends to zero with $\delta' \longrightarrow 0$.

The main formula (1.4) has now been proved without exception.

$$\int_{r_0}^{R} n_a(r) \cdot \frac{dr}{r} = [\psi_a(r)]_{r_0}^{R}$$

equals $\phi(a)$ where

$$(1.7)\quad \phi(z) = \phi(R;z) = \begin{cases} 0 & \text{for } |z| \geqq R, \\ \log R/|z| & \text{for } r_0 \leqq |z| \leqq R, \\ \log R/r_0 & \text{for } |z| \leqq r_0. \end{cases}$$

Hence we may write instead of (1.4):

$$(1.8)\quad \sum_a \phi(a) - \sum_b \phi(b) = \sum_z \nu(z)\phi(z) = \mathfrak{M}_r \log |f| \,]_{r_0}^{R},$$

where a runs over all zeros, b over all poles of f, each counted with its multiplicity. Indeed, the expression (1.3) for N(R) changes into the left side of (1.8) by substituting $\sum_{|z|<r} \nu(z)$ for n(r) and then interchanging summation and integration.

The function $\phi(z)$ is the electric potential of the circular condenser $r_0 \leqq |z| \leqq R$ whose inner layer of radius r_0 carries unit charge while the circle of radius R (carrying the opposite charge) is its outer layer. Our fundamental equation (1.8) will therefore be called the <u>condenser formula</u>. It is in this form (1.8) that it will carry over to arbitrary Riemann surfaces. [The name "argument principle" used by the Finnish school is proper for the equation (1.1) because $\Im \log f$ is the argument of the complex value f, but it is hardly adequate for the integrated equation (1.4) and certainly inadequate for its generalization to arbitrary Riemann surfaces.]

If f is rational, then n(r) will be, from a certain r on, the sum ν' of the orders of f at all finite points z and hence N(r) will differ from $\nu' \cdot \log r$ by a function of r which stays bounded as r approaches infinity. Thus N(r) is a substitute for that total order with the provision that two functions of r are considered equivalent in case their difference is bounded. We use the symbol ~ for the equivalence thus defined. In the general case of meromorphic functions and curves the "order" will of necessity be expressed by a function of r rather than by a number, equivalent functions defining the same order. But one more word about a rational f. Let it be of order $\mu = \nu_\infty$ at infinity. The Taylor expansion

$$f(z) = a_\mu (1/z)^\mu + \dots \qquad (a_\mu \neq 0)$$

around $z = \infty$ shows at once that

$$\log |f(re^{i\vartheta})| \sim -\mu \log r, \quad \mathfrak{M}_r \log |f| \sim -\mu \log r.$$

Hence the relation (1.4) gives indeed $\nu' = -\nu_\infty$ or $\nu' + \nu_\infty = 0$ as it should be.

For a fixed point a the potential $\phi(R;a)$ is a very simple function of $u = \log R$. Write

$$u_0 = \max (\log |a|, \log r_0),$$

then

$$\phi(R;a) = \begin{cases} 0 & \text{for } u \leqq u_0 \\ u - u_0 & \text{for } u \geqq u_0. \end{cases}$$

Thus $\phi(R;a)$ is a non-negative non-decreasing convex function of log R.

§2. The first main theorem

Cashing in on all the preparations in Chapters I and II we shall now derive the first decisive result about meromorphic curves.* A given non-degenerate meromorphic

*As mentioned in the preface, the systematic theory of meromorphic curves, along the lines of Nevanlinna's work on meromorphic functions, was inaugurated by H. and J. Weyl in 1938. But a number of older investigations, of which we were unaware at that time, pointed in the same direction; in addition to an important theorem by E. Borel, which will turn up here at the very end of our story, we mention: A. Bloch, Ann. de l'Ecole Norm. 43, 1926, 309. P. Montel, C. R. Acad. Sci., Paris 189, 1929, 625 and 731. R. Nevanlinna, Le théorème de Picard-Borel et le théorie des fonctions méromorphes, Paris 1929. H. Cartan, Thèse, Paris 1928, and Mathematica (Cluj) 7, 1933, 5.

curve associates with each value of z_0 a reduced representation

$$x_i = \mathfrak{P}_i(z-z_0) = c_i + c_i'(z-z_0) + \dots \quad (i=1,\dots,n)$$
$$(c_1, \dots, c_n) \neq (0, \dots, 0).$$

Let α, β be two non-vanishing contravariant vectors. Following the model of rational curves, we apply the condenser formula to the meromorphic function

$$f = \sum \alpha_i x_i / \sum \beta_i x_i = (\alpha x)/(\beta x).$$

$\sum \alpha_i \mathfrak{P}_i(z-z_0)$ will vanish for $z = z_0$ to a certain order $\nu(z_0;\alpha)$. We form

$$n(r;\alpha) = \sum_{|z|<r} \nu(z;\alpha),$$

$$N(R;\alpha) = \int_{r_0}^{R} n(r;\alpha)\,\frac{dr}{r} = \sum_z \nu(z;\alpha)\phi(z).$$

The order of f at $z = z_0$ equals $\nu(z_0;\alpha) - \nu(z_0;\beta)$; hence

$$(2.1) \quad N(R;\alpha) - N(R;\beta) = \left[\frac{1}{2\pi}\int_0^{2\pi} \log\left|\frac{(\alpha x)}{(\beta x)}\right| \cdot d\vartheta\right]_{r_0}^{R}.$$

It is understood, here as always when nothing else is said explicitly, that in an integral extending over ϑ the argument z of the integrand is to be replaced by $re^{i\vartheta}$. One is tempted to break the integrand into

$$\log|(\alpha x)| - \log|(\beta x)|,$$

thus to be led to the conclusion that

$$N(R;\alpha) + \left[\frac{1}{2\pi}\int_0^{2\pi} \log\frac{1}{|(\alpha x)|} \cdot d\vartheta\right]_{r_0}^{R}$$

has the same value for β as for α and is therefore independent of the intersecting plane $(\alpha x) = 0$.

However, this will not do, because only the ratios of the coordinates x_i have a meaning and are meromorphic function of z. We get around this difficulty by introducing the unitary metric connected with our homogeneous coordinates x_i in projective n-space according to which the vector x has a length $|x|$ defined by

$$|x|^2 = |x_1|^2 + \dots + |x_n|^2.$$

We replace $|(\alpha x)|$ by the point-plane distance

$$\|\alpha x\| = \frac{|(\alpha x)|}{|\alpha| \cdot |x|}$$

which indeed depends on the ratios of the α and the ratios of the x only. Formula (2.1) remains correct if we write $\frac{\|\alpha x\|}{\|\beta x\|}$ for $\frac{|(\alpha x)|}{|(\beta x)|}$, and then the splitting of

$$\log \frac{\|\alpha x\|}{\|\beta x\|} \text{ into } \log \|\alpha x\| - \log \|\beta x\|$$

is perfectly legitimate; and after introducing the compensating function or "defect" m by

$$(2.2) \qquad m(r;\alpha) = \frac{1}{2\pi}\int_0^{2\pi} \log \frac{1}{\|\alpha x\|} \cdot d\vartheta$$

we have proved the

FIRST MAIN THEOREM. The sum

$$(2.3) \qquad T(R) = N(R;\alpha) + [m(r;\alpha)]_{r_0}^{R}$$

is independent of α.

We call T(R) the order, or more precisely the order function, of our meromorphic curve. In the defining relation

$$(2.4) \qquad N(R;\alpha) + m(R;\alpha) = T(R) + m(r_0;\alpha)$$

we look upon R as a variable tending to infinity. What can we say about the behavior of the various terms?

The last term $m(r_0;\alpha)$ on the right is a constant with respect to R and hence of little importance. Since the distance $\|\alpha x\| \leq 1$, the compensating function $m(r;\alpha)$ is non-negative. From this fact of paramount importance there follows at once the inequality

$$N(R;\alpha) \leq T(R) + m(r_0;\alpha), \tag{2.5}$$

so that for no intersecting plane $(\alpha x) = 0$ the function $N(R;\alpha)$, which accounts for the intersections, grows more rapidly than the order $T(R)$. The quantity $N(R;\alpha)$ vanishes for $R = r_0$ and is a non-decreasing convex function of log R. Any function of R with these properties shall be said to be of regular type. Either of the two expressions

$$N(R;\alpha) = \int_{r_0}^{R} n(r;\alpha) d \log r = \sum_z \nu(z;\alpha) \cdot \phi(R;z)$$

may be used to prove that N is of regular type: In the first expression $n(r;\alpha)$ is a positive increasing function of r, in the second the coefficients $\nu(z;\alpha)$ are non-negative and independent of R while we have observed at the end of the last section that $\phi(R;z)$ is a non-decreasing convex function of log R.

If our curve is rational and $\nu'(\alpha)$, $\nu_\infty(\alpha)$ denote the numbers of intersections with the plane (α) at all finite spots and at the infinite spot $z = \infty$ respectively, then

$$N(R;\alpha) \sim \nu'(\alpha) \cdot \log R, \qquad m(R;\alpha) \sim \nu_\infty(\alpha) \cdot \log R,$$

hence

$$T(R) \sim \nu \log R$$

where $\nu = \nu'(\alpha) + \nu_\infty(\alpha)$ is the order of the rational curve, which indeed is independent of α. Thus we are well justified in calling m the compensating, and T the order, function.

All this looks pretty good. And yet we seem to have taken a step of the utmost gravity by introducing a unitary metric depending on the system of coordinates x_i. It has the effect that, contrary to the orders of rational and algebraic curves, the general order function T is not a projective invariant; it is invariant with respect to unitary, but not to arbitrary, linear transformations of the coordinates. Fortunately this circumstance is much less serious than at first appears. For we have seen in Chapter I, §5, that the distances $\|\alpha x\|$ measured in two different unitary metrics, which we shall now denote by $\|\alpha x\|$ and $\|\alpha x\|'$, have a ratio varying between constant positive bounds $\lambda \geqq 1$ and $1/\lambda$; we determined these bounds explicitly. Hence

$$- \log \lambda \leqq \log \frac{1}{\|\alpha x\|'} - \log \frac{1}{\|\alpha x\|} \leqq \log \lambda ,$$

$$- \log \lambda \leqq m'(r;\alpha) - m(r;\alpha) \leqq \log \lambda ,$$

$$|T'(R) - T(R)| \leqq 2 \log \lambda .$$

A change of the coordinate system therefore affects the order function T(R) only in an unessential manner, $T'(R) \sim T(R)$: <u>in the sense of equivalence</u> T(R) <u>is independent of the projective coordinates employed</u>. Two meromorphic curves $\mathfrak{C}$, $\mathfrak{C}_I$ are said to be <u>of the same order</u> if and only if their order functions T(R), $T_I(R)$ are <u>equivalent</u>. The substance of the equation (2.4) is more clearly exhibited by writing it as an equivalence:

$$N(R;\alpha) + m(R;\alpha) \sim T(R).$$

Any isolated set of points z_1, z_2, ... in the z-plane, like that of the intersections of a plane (α) with $\mathfrak{C}$, in which points are counted with certain multiplicities, is best characterized by the quantity $\nu(z)$ which gives the number of points in the set coinciding with z. Thus

$$\sum_{|z_i|<R} 1 = \sum_{|z|<R} \nu(z) = n(R)$$

is the number of points z_i inside the circle of radius R, while we propose to call

$$N(R) = \sum_i \phi(R;z_i) = \sum_z \phi(R;z)\nu(z)$$

the valence of these points in the same circle. N(R) is always a function of regular type. Number and valence are interrelated by the equation

$$N(R) = \int_{r_0}^{R} n(r)\,\frac{dr}{r}.$$

A theorem similar to that for $\mathfrak{C}$ holds for the associated $\mathfrak{C}_p$. With any two non-vanishing special contravariant p-ads A, B one forms the meromorphic function

$$(AX^p)/(BX^p)$$

the numerator and denominator of which are linear forms of the osculating p-element

$$X^p = [x, dx/dz, \ldots, d^{p-1}x/dz^{p-1}].$$

Let (AX^p) vanish for $z = z_0$ in the order $d_p(z_0) + \nu_p(z_0;A)$, $d_p(z_0)$ being the multiplicity of the common zero of all components $X^p(i_1,\ldots,i_p)$ at $z = z_0$.

$$(2.6) \qquad N_p(R;A) = \sum_z \nu_p(z;A)\phi(R;z) \qquad \text{for } A = {*A^{n-p}}$$

may be described as the valence in the circle of radius R of the intersections of a fixed (n-p)-element A^{n-p} with the osculating p-elements of the curve. Setting

$$(2.6') \quad m_p(r;A) = \frac{1}{2\pi}\int_0^{2\pi} \log \frac{1}{\|AX^p\|}\cdot d\vartheta = \frac{1}{2\pi}\int_0^{2\pi} \log \frac{1}{\|A^{n-p}:X^p\|}\cdot d\vartheta$$

$$(z = re^{i\vartheta})$$

we obtain the

FIRST MAIN THEOREM FOR ARBITRARY RANK p.

$$N_p(R;A) + [m_p(r;A)]_{r_0}^{R} = T_p(R) \tag{2.7}$$

is independent of A .

Again m_p is non-negative. $T_p(R)$ is the order function of rank p. By changing the underlying unitary metric $m_p(r;A)$ changes by an additive function which keeps within the fixed bounds $\pm \log \lambda_p$.

$\nu_p(z_0;A)$ becomes the vanishing order of $(A\,X^p)$ at $z = z_0$ only after removal of the fixed (A-independent) zero of order $d_p(z_0)$. One might therefore describe $\nu_p(z_0;A)$ as the multiplicity of the zero of

$$\frac{|(A\,X^p)|}{|A| \cdot |X^p|} = \|A\,X^p\|$$

at $z = z_0$, and in this sense we use for $n_p(r;A)$ and $N_p(R;A)$ the more explicit notations

$$n(r, \|A\,X^p\|), \qquad N(R, \|A\,X^p\|)$$

which exhibit the quantity the zeros of which are counted. The expression of N is thereby brought into closer analogy with that of m.

At the beginning we have made the assumption that our curve is non-degenerate. But the results still hold in the degenerate case if one simply excludes the "critical" vectors α or p-ads A for which (αx), $(A\,X^p)$ vanish identically along the given curve.

The orders T_p, T_p^* of a curve $\mathfrak{C}$ and its dual $\mathfrak{C}^*$ are obviously connected by the relations

$$T_p^* = T_{n-p}.$$

The whole argument as presented here will later carry over to analytic curves stemming from an arbitrary Riemann surface. But in our special case we can use

Weierstrass's construction and assume the curve to be given in the normalized global form

$$x_i = f_i(z) \tag{2.8}$$

where f_i are entire functions without common zeros. We normalize $|\alpha| = 1$, $|A| = 1$ and understand by (αx), (AX^p) the entire functions of z which arise from these linear forms of x and X^p respectively by the substitution (2.8). Because $\nu(z_0;\alpha)$ is now the multiplicity of the zero of (αx) at $z = z_0$ we find by application of the condenser formula to the entire function (αx):

$$N(R;\alpha) + \left[\frac{1}{2\pi}\int_0^{2\pi} \log \frac{1}{|(\alpha x)|}\cdot d\vartheta\right]_{r_0}^{R} = 0$$

or

$$N(R;\alpha) + [m(r,\alpha)]_{r_0}^{R} = \left[\frac{1}{2\pi}\int_0^{2\pi} \log |x|\cdot d\vartheta\right]_{r_0}^{R}.$$

Thus it has again been proved that the left member is independent of α and at the same time we obtain the formula

$$T(R) = \left[\frac{1}{2\pi}\int_0^{2\pi} \log |x|\cdot d\vartheta\right]_{r_0}^{R} \tag{2.9}$$

for the order function. One could even think of defining T by this simple equation. In our normalized global representation a factor of the form $e^{\kappa(z)}$, $\kappa(z)$ entire, is arbitrary. Putting this factor in adds the term $\Re\kappa(z)$ to $\log |x|$, but since the circular mean value

$$\frac{1}{2\pi}\int_0^{2\pi} \Re\kappa(z)\cdot d\vartheta \qquad (z=re^{i\vartheta})$$

of the harmonic function $\Re\kappa(z)$ is independent of the radius r, the function (2.9) is not affected by this change.

For higher rank p one has to observe that $\nu_p(z_0;A)$ is the multiplicity of the zero at $z = z_0$ of $|(AX^p)|/|X^p|$. Hence the condenser formula for the entire function

(AX^p),

$$N(R, |(AX^p)|) + \left[\frac{1}{2\pi}\int_0^{2\pi} \log \frac{1}{|(AX^p)|} \cdot d\vartheta\right]_{r_0}^R = 0,$$

yields

$$N\left(R, \frac{|(AX^p)|}{|X^p|}\right) + [m_p(r;A)]_{r_0}^R =$$

$$-N(R, |X^p|) + \left[\frac{1}{2\pi}\int_0^{2\pi} \log |X^p| \cdot d\vartheta\right]_{r_0}^R .$$

We thus arrive at the following expression for $T_p(R)$:

$$(2.10) \qquad T_p(R) = \left[\frac{1}{2\pi}\int_0^{2\pi} \log |X^p| \cdot d\vartheta\right]_{r_0}^R - D_p(R)$$

where $D_p(R) = N(R, |X^p|)$ is defined by

$$(2.11) \qquad D_p(R) = \sum_z d_p(z) \cdot \phi(R;z).$$

§3. Meromorphic functions. The exponential function

A meromorphic function f may be conceived as a meromorphic curve in 2-space, $f = x_1/x_2$. Denoting its order function by $T\{f\} = T(R;f)$ we have the relation

$$T\left\{\frac{\alpha f+\beta}{\gamma f+\delta}\right\} \sim T\{f\}$$

for any non-singular linear transformation with constant coefficients α, β, γ, δ and

$$T\left\{\frac{\alpha f+\beta}{\gamma f+\delta}\right\} = T\{f\}$$

if that transformation is unitary. In particular,

$$T\{1/f\} = T\{f\} \quad \text{and} \quad T\{f-a\} \sim T\{f\}.$$

The last equation connects the investigation of the a-places of f with that of the zeros.

By applying our first main theorem to the values $\alpha_1 = 1$, $\alpha_2 = 0$ and $\alpha_1 = 0$, $\alpha_2 = 1$ respectively we obtain

$$(3.1)\quad N_0(r) + \frac{1}{2\pi}\int_0^{2\pi} \log \sqrt{(1+|1/f|^2)} \cdot d\vartheta = T(r) + \text{const.},$$

$$(3.1')\quad N_\infty(r) + \frac{1}{2\pi}\int_0^{2\pi} \log \sqrt{(1+|f|^2)} \cdot d\vartheta = T(r) + \text{const.}$$

where $N_0(r)$, $N_\infty(r)$ denote the valences of the zeros of f in the circle K_r of radius r. (The coincidence of the two expressions (3.1) and (3.1') is nothing else but the law $T\{1/f\} = T\{f\}$.) For $\alpha_1 = 1$, $\alpha_2 = -a$ we find

$$(3.2)\quad N_a(r) + \frac{1}{2\pi}\int_0^{2\pi} \log \frac{\sqrt{(1+|f|^2)}}{|f-a|} \cdot d\vartheta = T(r) + \text{const.}$$

where $N_a(r)$ is the valence in K_r of the a-places of f.

The equations (3.1) are due to R. Nevanlinna and constitute the first main theorem of his epoch-making theory of meromorphic functions. In his paper they appear in a slightly different form. Instead of the unitary metric with the "circle"

$$|x|_c^2 = |x_1|^2 + |x_2|^2 = 1$$

as gauge figure he uses the "square" (or, as it is sometimes called, the "dicylinder")

$$|x|_s = \max(|x_1|, |x_2|) = 1$$

and thus gets another order function T_s instead of our $T = T_c$. The difference, however, is negligible because of the evident inequalities

$$|x|_s^2 \leqq |x|_c^2 \leqq 2 \cdot |x|_s^2,$$

$$0 \leqq \log |x|_c - \log |x|_s \leqq \tfrac{1}{2} \log 2,$$

which imply

$$0 \leqq T_c(r) - T_s(r) \leqq \log 2, \qquad T_c \sim T_s.$$

Setting

$$\log^+ c = \log c \text{ if } c \geqq 1, \quad = 0 \text{ if } c \leqq 1$$

for any positive number c, one has

$$\log \max (c,1) = \log^+ c.$$

Hence Nevanlinna's formulas

$$(3.3)\qquad \begin{aligned} N_0(r) + \frac{1}{2\pi}\int_0^{2\pi} \log^+ |1/f| \cdot d\vartheta &= T_s(r) + \text{const.}, \\ N_\infty(r) + \frac{1}{2\pi}\int_0^{2\pi} \log^+ |f| \cdot d\vartheta &= T_s(r) + \text{const.} \end{aligned}$$

The observation is forced upon us that in the sense of equivalence the order function T is not influenced by the metric even if we admit metrics based on arbitrary gauge figures (Minkowski metrics) rather than on ellipsoids. Only later will the advantage of the unitary metrics become apparent.

For an <u>entire</u> function f the equation (3.1') gives the definition of the order function T(r) in terms of the growth of f, namely

$$(3.4)\qquad T(r) = \frac{1}{2\pi}\int_0^{2\pi} \log \sqrt{1+|f|^2} \cdot d\vartheta \qquad (z=re^{i\vartheta}).$$

The same formula follows from (2.9). The older theory used instead

$$\log M(r) = \log \max_{0 \leqq \vartheta \leqq 2\pi} |f(re^{i\vartheta})|.$$

It is clear that

$$T(r) \leqq \log \sqrt{1+M^2(r)}$$

or essentially $T(r) \leqq \log M(r)$. The first equation (3.1) connects the distribution of zeros with the growth of f. <u>Under all circumstances we have</u>

$$(3.5)\qquad N_0(r) \leqq T(r) + \text{const.}$$

However, besides the distribution of zeros, a certain compensating function enters into the formula, allowing $N_0(r)$ to fall behind the "normal" value $T(r)$.

The <u>exponential</u> <u>function</u> is an example to the point. With Nevanlinna's square metric we find

$$T_s(r) = \frac{1}{2\pi} \int_0^{2\pi} \log \max (1, e^{r \cdot \cos \vartheta}) \cdot d\vartheta$$

$$= \frac{1}{2\pi} \int_{-\pi/2}^{+\pi/2} r \cdot \cos \vartheta \cdot d\vartheta = r/\pi,$$

whereas with our definition (3.4) this equation may not be exactly true, but the corresponding equivalence

$$T(r) \sim r/\pi$$

is correct. On the other hand

$$M(r) = e^r, \quad \log M(r) = r.$$

The exponential function has no zeros; and indeed the compensating function in (3.1)

$$\frac{1}{2\pi} \int_0^{2\pi} \log \sqrt{1+|f|^{-2}} \cdot d\vartheta = \frac{1}{2\pi} \int_0^{2\pi} \log \sqrt{1+e^{-2r \cos \vartheta}} \cdot d\vartheta$$

is identical with

$$\frac{1}{2\pi} \int_0^{2\pi} \log \sqrt{1+|f|^2} \cdot d\vartheta = \frac{1}{2\pi} \int_0^{2\pi} \log \sqrt{1+e^{2r \cos \vartheta}} \cdot d\vartheta,$$

as the substitution $\vartheta \rightarrow \pi + \vartheta$ proves. But a being any value $\neq 0, \infty$, the equation $e^z = a$ has solutions which are distributed with equal distance 2π on a vertical line, $z = z_0 + 2h\pi i$ $(h=0,\pm 1,\pm 2,\ldots)$. Hence the number $n_a(r)$ of solutions within the circle $|z| < r$ equals $r/\pi + O(1)$ and consequently

(3.6) $$N_a(r) = r/\pi + O(\log r).$$

On the other hand a direct estimate which will be carried out under somewhat more general circumstances in §5 will show the compensating term

$$m_a(r) = \frac{1}{2\pi} \int_0^{2\pi} \log \frac{\sqrt{1+|e^z|^2}}{|e^z-a|} \cdot d\vartheta$$

to be ~ 0, and thus the relation (3.2) will be verified. For the values $a = 0, \infty$ the N-part vanishes and the compensating term tells the whole story, whereas for $a \neq 0, \infty$ the compensating term is negligible and the N-part is the only one that counts. These conditions, especially equation (3.6), clearly reveal the superiority of Nevanlinna's order function $T(r)$ which is equivalent to r/π for the exponential function, over the older function $\log M(r)$ in whose asymptotic law $\log M(r) \sim r$ the factor $1/\pi$ is missing.

We have now reached the point where our theory ties up with the early history as sketched in the introduction. Let $f(z)$ be an entire function and $a_1, a_2, \ldots$ its zeros counted with their multiplicity and arranged by increasing modulus, $|a_1| \leqq |a_2| \leqq \ldots$. We repeat Hadamard's first theorem: For any $\lambda > 0$ convergence of the integral

$$\int^{\infty} \log M(r) \cdot r^{-(\lambda+1)} dr$$

implies convergence of the sum $\sum' |a_\nu|^{-\lambda}$ (the accent indicating that initial members for which $a_\nu = 0$ are to be omitted). The connection with $N_0(r)$ is established by the following elementary

LEMMA 3. A.

$$\sum{}' |a\ |^{-\lambda}, \int^{\infty} n_0(r) \cdot r^{-(\lambda+1)} dr, \int^{\infty} N_0(r) \cdot r^{-(\lambda+1)} dr$$

converge and diverge simultaneously.

Proof. Omitting the subscript 0 we use the equation

(3.7) $$\int_{r_0}^{r} \frac{dn(r)}{r^\lambda} = \left\{\frac{n(r)}{r^\lambda} - \frac{n(r_0)}{r_0^\lambda}\right\} + \lambda \int_{r_0}^{r} \frac{n(r)dr}{r^{\lambda+1}} .$$

If $\sum' |a_\nu|^{-\lambda}$ converges, i. e. if the left side is bounded, so is

(3.8) $$\int_{r_0}^{r} \frac{n(r)dr}{r^{\lambda+1}} .$$

<u>Vice versa</u>

$$\lambda \int_r^{2r} n(r) \cdot r^{-(\lambda+1)}dr \geqq n(r) \cdot \lambda \int_r^{2r} r^{-(\lambda+1)}dr = \frac{n(r)}{r^\lambda}(1-2^{-\lambda}).$$

Convergence of (3.8) for $r \longrightarrow \infty$ therefore implies

(3.9) $$n(r)/r^\lambda \longrightarrow 0$$

and then equation (3.7) proves that its left member is also convergent. In the same fashion one passes from

$$\int \frac{dN(r)}{r^\lambda} = \int \frac{n(r)dr}{r^{\lambda+1}} \text{ to } \int \frac{N(r)dr}{r^{\lambda+1}} .$$

After having established the lemma we may modify Hadamard's statement as follows: Convergence of $\int^\infty r^{-(\lambda+1)} \cdot \log M(r) \cdot dr$ implies that of $\int^\infty r^{-(\lambda+1)} N_0(r)dr$. The true reason for this is now revealed by Nevanlinna's simple inequality (3.5) which obviously amounts to a greatly sharpened form of Hadamard's theorem.

§4. A fundamental lemma

Before going on with the theory of meromorphic curves, we develop a function-theoretical lemma useful for the discussion of the compensating term. Let $f_0(t)$ be a function defined and regular analytic in a certain region $\mathfrak{G}$ of the t-plane surrounding the segment $\mathfrak{g}$: $-c \leqq t \leqq c$ on the real axis. We assume that $f_0(t)$ does not vanish identically.

LEMMA 4. A. Under these hypotheses there exist two numbers $b > 0$ and B such that

(4.1) $$L\{f\} = \frac{1}{2c}\int_{-c}^{+c} \log |f(t)| \cdot dt \geqq -B$$

for every regular function f(t) in $\mathfrak{G}$ which satisfies the inequality

(4.2) $$|f(t) - f_0(t)| \leqq b \qquad \text{throughout } \mathfrak{G}.$$

Proof. Suppose that $f_0(t)$ has h zeros on $\mathfrak{g}$. We can surround $\mathfrak{g}$ with a closed rectangular strip Q inside $\mathfrak{G}$,

$$|\Re t| \leqq c + \delta, \qquad |\Im t| \leqq \delta,$$

in which $f_0(t)$ has no other zeros. On the boundary Q' of Q the modulus $|f_0(t)|$ will then have a positive lower bound 2b, and $|f(t)| \geqq b$ on Q' for every f(t) satisfying (4.2). Moreover f(t) possesses the same number h of zeros inside Q as $f_0(t)$ does. Indeed, consider the function

$$f_\lambda(t) = f_0(t) + \lambda(f(t) - f_0(t))$$

involving a real parameter λ varying from 0 to 1 and observe that $|f_\lambda(t)| \geqq b$ on Q'. The number of the zeros of $f_\lambda(t)$ in Q is given by the contour integral of its logarithmic derivative

$$\frac{1}{2\pi i}\int_{Q'}\left(\frac{df_\lambda}{dt}\Big/ f_\lambda\right)\cdot dt.$$

Because the denominator of the integrand is bounded away from zero, the integral itself varies continuously while λ travels from 0 to 1 and, always being an integer, must be a constant.

Designate by d the length of the diagonal of Q and by $t_1, \ldots, t_h$ the zeros of f(t); then form the function

$$\frac{(t-t_1)\ldots(t-t_h)}{f(t)}$$

which is regular throughout Q and is of modulus $\leqq d^h/b$ on the boundary Q'. Hence the inequality

$$\left|\frac{(t-t_1)\ \dots\ (t-t_h)}{f(t)}\right| \leqq d^h/b$$

holds also inside Q, in particular along the segment g, or

$$(4.3)\qquad |f(t)| \geqq b\cdot\frac{|t-t_1|\ \dots\ |t-t_h|}{d^h},$$

and therefore

$$-L\{f\} \leqq \log 1/b + \sum_{j=1}^{h}\frac{1}{2c}\int_{-c}^{+c}\log\frac{d}{|t-t_j|}\cdot dt.$$

One readily sees (the proof will presently be given in a generalized form) that the integral

$$\frac{1}{2c}\int_{-c}^{+c}\log\frac{d}{|t-t_0|}\cdot dt$$

takes on its maximum value for $t_0 = 0$, which is $1 + \log d/c$. Thus

$$-L\{f\} \leqq B = \log 1/b + h(1+\log d/c).$$

How this lemma may be utilized for our purposes is shown by

LEMMA 4. B. For a fixed radius $r = r_0$ the compensating term $m(r_0;\xi)$ of a non-degenerate curve $\mathfrak{C}$ has an upper bound for all $\xi \neq 0$.

Proof. We use a representation $x_i = x_i(z)$ of $\mathfrak{C}$ by functions $x_i(z)$ which are regular in a circle $|z| < R$ of radius R larger than r_0; this normalization of the gauge factor is the trivial one mentioned in Chap. I, §11, which does not resort to Weierstrass's construction. The functions

$$x_i(r_0e^{it}) = f_i(t)$$

are regular in the half plane $\Im t < \log R/r_0$. Take any fixed contravariant vector α of length 1 and by means of it and n parameters ξ_k form the two functions

$$f_0(t) = \sum \alpha_k f_k(t), \qquad f(t) = \sum \xi_k f_k(t).$$

Our lemma (with $c = \pi$) yields a neighborhood $\mathfrak{N}_\alpha$ of α on the unit sphere,

$$|\xi - \alpha| < \rho, \qquad |\xi| = 1,$$

and a constant upper bound B_α of $m(r_0; \xi)$ valid for all $\xi \in \mathfrak{N}_\alpha$. Because the sphere $|\xi| = 1$ is compact, it can be covered by a finite number of such neighborhoods $\mathfrak{N}_\alpha$, and the largest of the B_α corresponding to them gives the desired universal bound for $m(r_0; \xi)$. -- Instead of the coverage argument one might use one of the other more or less equivalent general principles of analysis; but an appeal to one of these principles, which are so suspicious to the intuitionist, seems inevitable to take us from local to universal boundedness.

For the investigation of exponential curves we need a slight generalization of Lemma B. Suppose we are given m linearly independent functions $f_k(t)$ regular in $\mathfrak{G}$ and m further functions $f_k(\epsilon; t)$, also regular in $\mathfrak{G}$ but depending, besides on t, on a positive parameter ϵ in such a way that

$$f_k(\epsilon; t) \longrightarrow f_k(t) \text{ with } \epsilon \longrightarrow 0 \text{ uniformly for } t \in \mathfrak{G}.$$

Let $l_1, \ldots, l_m$ be non-negative numbers, l_1 actually positive.

LEMMA 4. C. Under the hypotheses just enumerated there exist numbers $\epsilon_0 > 0$ and B such that

$$(4.4) \quad \frac{1}{2c} \int_{-c}^{c} \log |\xi_1 f_1(\epsilon; t) + \ldots + \xi_m f_m(\epsilon; t)| \, dt \geq -B$$

for all ξ for which

$$|\xi_1| = l_1 > 0, \qquad |\xi_2| = l_2, \; \ldots \;, \qquad |\xi_m| = l_m$$

(torus) and all positive $\epsilon \leqq \epsilon_0$.

Proceed exactly as before: For any $\xi = \alpha$ on the torus, Lemma A with

$$f_0(t) = \sum \alpha_k f_k(t), \qquad f(t) = \sum \xi_k f_k(\epsilon;t)$$

yields two numbers $\epsilon_\alpha > 0$ and B_α and a neighborhood $\mathfrak{N}_\alpha$ of α such that the left member of (4.4) has the lower bound $-B_\alpha$ for $0 < \epsilon < \epsilon_\alpha$ and $\xi \in \mathfrak{N}_\alpha$. Then use compactness of torus.

We return to Lemma A in order to prove not only boundedness but a certain property of uniform continuity for the integral (4.1). But now we find it convenient to bring that integral into somewhat closer accord with the one defining $m(r_0;\xi)$. Let $F_0(t)$ be a given positive continuous function on $\mathfrak{g}$. For functions $f(t)$ which are regular analytic in $\mathfrak{G}$ and majorized by $F_0(t)$ on $\mathfrak{g}$,

$$|f(t)| \leqq F_0(t) \quad \text{for} \quad -c \leqq t \leqq c,$$

(admissible functions) we consider the integral

$$J\{f\} = \int_{-c}^{+c} \log \frac{F_0(t)}{|f(t)|} \cdot dt$$

and compare it with the integral $J_\omega\{f\}$ of the same non-negative integrand extending over that part of $\mathfrak{g}$ where the integrand does not exceed a given constant ω.

LEMMA 4. D. Given $F_0(t)$ and an admissible function $f_0(t) \not\equiv 0$, there exists a positive constant b such that no admissible function $f(t)$ satisfying the inequality

$$|f(t) - f_0(t)| \leqq b \quad \text{in } \mathfrak{G}$$

vanishes identically and that $J_\omega\{f\}$ tends to $J\{f\}$

with $\omega \longrightarrow \infty$ uniformly for all those functions f(t).

Proof. Use the same construction of the rectangle Q and the number b as in the proof of Lemma A, and let A be an upper bound of $F_0(t)$ on $\mathfrak{q}$. Choose a positive η and exclude from $\mathfrak{q}$ the set (ω) of all points t which satisfy any of the h inequalities

$$|t-t_j| \leqq \eta/h \qquad (j=1,\ldots,h).$$

On the complement $\mathfrak{q} - (\omega)$ the inequality (4.3) implies

$$\frac{F_0(t)}{|f(t)|} \leqq \frac{A}{b}\left(\frac{\eta}{hd}\right)^{-h},$$

hence if η is determined by the equation

$$\log \left\{\frac{A}{b}\left(\frac{\eta}{hd}\right)^{-h}\right\} = \omega, \qquad \log \frac{\eta}{d} = \log h + \frac{1}{h}\left(\log \frac{A}{b} - \omega\right)$$

we get

$$(4.5) \qquad 0 \leqq J\{f\} - J_\omega\{f\} \leqq \int_{(\omega)} \log \frac{F_0(t)}{|f(t)|} \cdot dt.$$

The set (ω) is the join of a number $\leqq h$ of non-overlapping intervals of total length $\leqq 2\eta$. It is reasonable to assume ω already so big, $\omega \geqq \omega_0$, as to make $\eta \leqq c$. In the right member of (4.5) we use the estimate

$$\log \frac{F_0(t)}{|f(t)|} \leqq \log \frac{A}{b} + \sum_{j=1}^{h} \log \frac{d}{|t-t_j|}.$$

But, as we shall presently see, the integral

$$(4.6) \qquad \int_{(\omega)} \log \frac{d}{|t-t_0|} \cdot dt$$

extending over any set of intervals (ω) of given total length $2\eta' \leqq 2\eta$ takes on its largest value

$$(4.7) \qquad \int_{-\eta'}^{\eta'} \log \frac{d}{|t|} \cdot dt$$

if (ω) is a single interval of length $2\eta'$ with its center in t_0. Because of $\eta \leqq d$, (4.7) is further increased by extending the limits of integration to $\pm\eta$, and we thus arrive at the inequality

$$J\{f\} - J_\omega\{f\} \leqq 2\eta(\log \frac{A}{b}+1+\log d/\eta)$$

proving our lemma.

Should t_0 be off the real axis, the integral (4.6) obviously increases by dropping t_0 on the real axis:

$$|t-t_0| \geqq |t-\Re t_0| \quad \text{if } t \text{ is real.}$$

Hence we may assume t_0 to be real or, without loss of generality, $t_0 = 0$. Think of the individual intervals composing (ω) as rigid rods movable along the real axis. The integral is further increased by drawing these rods together towards the origin as closely as we can bring them without causing them to overlap. We then obtain a single interval $-\eta_0 \leqq t \leqq \eta_1$ containing the origin. If it is asymmetric with respect to the origin, if e. g. $\eta_1 > \eta_0$, we cancel the part extending from $\frac{1}{2}(\eta_0+\eta_1)$ to η_1 and replace it by the integral from $-\eta_0$ to $-\frac{1}{2}(\eta_0+\eta_1)$ which is the same as the integral from η_0 to $\frac{1}{2}(\eta_0+\eta_1)$; but this interval arises from the one it replaces by the displacement $-\frac{1}{2}(\eta_1-\eta_0)$ toward the origin. In this way we obtain the upper bound (4.7) for (4.6).

Application to the compensating function is obvious. Form the integral

$$m_\delta(r;\alpha) = \frac{1}{2\pi}\int \log \frac{1}{\|\alpha x\|} \cdot d\vartheta$$

extending only over that part of the full circle $-\pi \leqq \vartheta \leqq \pi$ where $\|\alpha x\| \geqq \delta$ (>0).

LEMMA 4. E. For a given non-degenerate curve $\mathfrak{C}$ and radius $r = r_0$ the integral $m_\delta(r_0;\xi)$ tends to $m(r_0;\xi)$ with $\delta \to 0$ uniformly for all $\xi \neq 0$.

The proof of course again appeals to the compactness argument for the sphere $|\xi| = 1$.

COROLLARY: For a fixed $r = r_0$ the defect $m(r_0;\alpha)$ is a continuous function of the intersecting plane (α).

Indeed, for any $\delta > 0$, the approximating function $m_\delta(r;\alpha)$ is a continuous function of (α).

All this is true provided the curve is non-degenerate. In the degenerate case we may assume that $x_1, \ldots, x_h$ are linearly independent, whereas $x_{h+1}, \ldots, x_n$ vanish identically. Then the defect $'m(r_0;\alpha)$ computed in the h-space with the coordinates $x_1, \ldots, x_h$ is continuous, except for $\alpha = 0$, but the true $m(r_0;\alpha)$ computed in the full space is connected with $'m(r_0;\alpha)$ by the equation

$$m(r_0;\alpha) - 'm(r;\alpha) = \frac{1}{2} \log \frac{|\alpha_1|^2 + \ldots + |\alpha_n|^2}{|\alpha_1|^2 + \ldots + |\alpha_h|^2} .$$

Hence $m(r_0;\alpha)$ loses its boundedness and continuity at all points α in the dual space, whose first h coordinates vanish.

§5. Exponential curves

A highly instructive illustration of the theory of meromorphic curves is provided by the exponential curves.* Given any n (complex) distinct numbers λ_i we define the curve by the equations

$$(5.1) \qquad x_1 = e^{\lambda_1 z}, \ldots, x_n = e^{\lambda_n z}.$$

The curve is non-degenerate because its Wronskian

*About the zeros of exponential sums cf. G. Pólya, Sitzungsber. Bayer. Ak. d. Wiss. 1920, 285-290, and E. Schwengeler, Geometrisches über die Verteilung der Nullstellen spezieller ganzer Funktionen, Thesis, Zürich, Switzerland, 1925.

$$|1, \lambda_k, \ldots, \lambda_k^{n-1}| \quad e^{(\lambda_1+\ldots+\lambda_n)z}$$

does not vanish identically; it even differs from zero for any value z.

First let us compute the order of our exponential curve which after dropping an additive constant is given by

$$T(r) = \frac{1}{2\pi}\int_0^{2\pi} \log |x| \cdot d\vartheta,$$

or in Nevanlinna's modified definition by

$$T_s(r) = \frac{1}{2\pi}\int_0^{2\pi} \log |x|_s \cdot d\vartheta, \quad |x|_s = \max(|x_1|, \ldots, |x_n|).$$

T(r) is equivalent to $T_s(r)$ because of

$$|x|_s^2 \leqq |x|^2 \leqq n \cdot |x|_s^2, \quad 0 \leqq \log |x| - \log |x|_s \leqq \frac{1}{2}\log n,$$

and it is easier to compute $T_s(r)$. Indeed, write

$$\lambda_k = a_k - ib_k \ (a_k, b_k \text{ real}), \quad z = x + iy = re^{i\vartheta},$$

then

$$\log |x_k| = a_k x + b_k y = r(a_k \cos \vartheta + b_k \sin \vartheta),$$

and setting

$$h(\vartheta) = \max(a_1 \cos \vartheta + b_1 \sin \vartheta, \ldots, a_n \cos \vartheta + b_n \sin \vartheta),$$

$$\int_0^{2\pi} h(\vartheta)\, d\vartheta = L \tag{5.2}$$

we find $T_s(r) = \frac{L}{2\pi} \cdot r$ and thus

$$T(r) \sim \frac{L}{2\pi} \cdot r. \tag{5.3}$$

The increase of T proportional to r is the most remarkable feature of this law. The coefficient L has a simple geometric significance.

We plot in a ξ,η-plane the n points $P_k : \xi = a_k$, $\eta = b_k$ and maintain that L is the circumference of the convex polygon $\mathfrak{D}$ spanned around the n points $P_1, \ldots, P_n$.

Take any index k. The n - 1 half-planes

$$(a_k - a_i)x + (b_k - k_i)y > 0 \qquad (i \neq k),$$

if they intersect at all, have a certain open angle Θ_k, $\vartheta_k < \vartheta < \vartheta_k'$, in common. Let us call k of the first or second kind according to whether this angle of intersection does or does not exist. For directions ϑ in the open Θ_k the kth of the quantities $P_i(\vartheta) = a_i \cos \vartheta + b_i \sin \vartheta$ is actually larger, in the closed angle $\vartheta_k \leqq \vartheta \leqq \vartheta_k'$ not smaller, than the others. Hence the full rose is divided into a number of angles in each of which a different one of the $P_k(\vartheta)$ rides on top. We label them

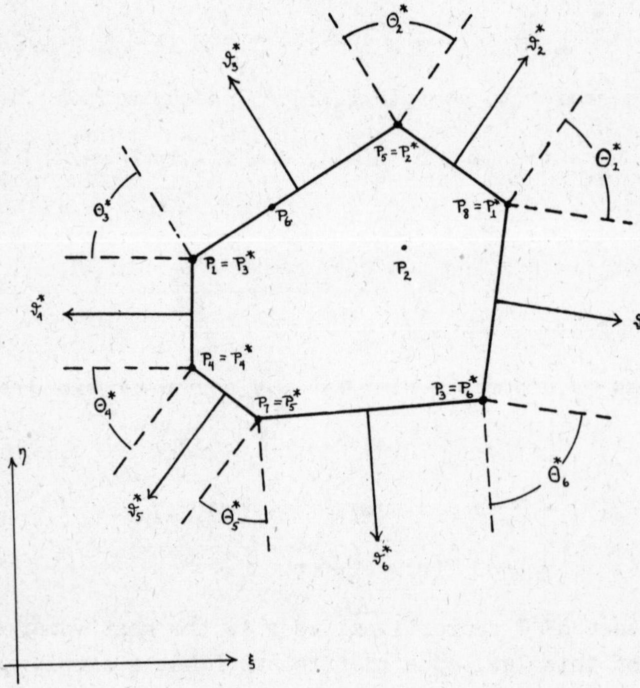

FIG. 1

as Θ_μ^* ($\mu = 1,\dots,\nu$) with the dividing directions ϑ_μ^*, as they follow one another around the rose. (The index μ ranges over the integers modulo a certain $\nu \leqq n$ in the cyclic order in which μ is followed by $\mu + 1$; the angle Θ_μ^* is bounded by the directions ϑ_μ^* and $\vartheta_{\mu+1}^*$, or ϑ_μ^* separates $\Theta_{\mu-1}^*$ and Θ_μ^*.) Let $k = k_\mu$ indicate the largest of the $P_k(\vartheta)$ in Θ_k, $\Theta_\mu^* = \Theta_{k_\mu}$. The points $P_\mu^* = P_k$ are the consecutive vertices of a polygon $\mathfrak{D}$ which deserves to be called the convex polygon surrounding the points $P_1, \dots, P_n$ of our diagram, for the following reason: the line joining the two points $P_{\mu-1}^*$ and P_μ^* of the diagram,

$$\xi \cos \vartheta_\mu^* + \eta \sin \vartheta_\mu^* = h(\vartheta_\mu^*),$$

leaves all points of the diagram on one side,

$$\xi \cos \vartheta_\mu^* + \eta \sin \vartheta_\mu^* \leqq h(\vartheta_\mu^*) \qquad \text{for } (\xi,\eta) = (a_k, b_k).$$

The contribution of the angle Θ_k to the integral (5.2) amounts to

$$\int_{\vartheta_k}^{\vartheta_k'} (a_k \cos \vartheta + b_k \sin \vartheta)\, d\vartheta =$$

$$a_k(\sin \vartheta_k' - \sin \vartheta_k) - b_k(\cos \vartheta_k' - \cos \vartheta_k),$$

therefore

$$\text{(5.4)} \quad \begin{aligned} L &= \sum_\mu \{a_\mu^*(\sin \vartheta_{\mu+1}^* - \sin \vartheta_\mu^*) - b_\mu^*(\cos \vartheta_{\mu+1}^* - \cos \vartheta_\mu^*)\} \\ &= \sum_\mu \{-(a_\mu^* - a_{\mu-1}^*) \sin \vartheta_\mu^* + (b_\mu^* - b_{\mu-1}^*) \cos \vartheta_\mu^*\}. \end{aligned}$$

If ϑ passes through ϑ_μ^* in positive sense, the sign of the expression

$$(a_\mu^* \cos \vartheta + b_\mu^* \sin \vartheta) - (a_{\mu-1}^* \cos \vartheta + b_{\mu-1}^* \sin \vartheta)$$

changes from - to +; therefore

$$a_\mu^* - a_{\mu-1}^* = -l_\mu \sin \vartheta_\mu^*, \quad b_\mu^* - b_{\mu-1}^* = l_\mu \cos \vartheta_\mu^*$$

and $l_\mu \sin(\vartheta - \vartheta_\mu^*)$ changes from - to +. Thus the factor l_μ is the <u>positive</u> length of the side $P_{\mu-1}^* P_\mu^*$, and equation (5.4) turns into $L = \sum l_\mu$. To round out the picture it ought to be mentioned that whatever the direction ϑ, all points P_k lie in the half plane $\xi \cos \vartheta + \eta \sin \vartheta \leqq h(\vartheta)$ and at least one on its boundary, the "supporting line of normal ϑ". For this reason $h(\vartheta)$ is called <u>function of support</u>.

In §3 we discussed the special case $n = 2$, $\lambda_1 = 0$, $\lambda_2 = 1$; the length of the surrounding "dygon" is $L = 2$.

Now let us cut the exponential curve (5.1) with an arbitrary plane $(\alpha x) = \sum \alpha_i x_i = 0$. In the special case just mentioned we were able to determine explicitly the points of intersection and thus derive an approximate value for $N(R;\alpha)$. Here we have to use the formula

$$N(R;\alpha) = \left[\frac{1}{2\pi} \int_0^{2\pi} \log |(\alpha x)| \cdot d\vartheta\right]_{r_0}^{R}$$

and will prove that

$$(5.5) \qquad \frac{1}{2\pi} \int_0^{2\pi} \log |(\alpha x)| \cdot d\vartheta \sim \frac{L}{2\pi} \cdot r$$

<u>provided all</u> n <u>coefficients</u> $\alpha_i \neq 0$. Hence in that case $N(R;\alpha)$ makes up the whole of $T(R)$ and there is no defect,

$$(5.6) \qquad m(r;\alpha) = \frac{1}{2\pi} \int_0^{2\pi} \log \frac{|\alpha| \cdot |x|}{|(\alpha x)|} d\vartheta \sim 0.$$

But we can also find out what happens in the exceptional cases where some of the α_i, though not all, vanish. Suppose for instance that the first m among them, $\alpha_1, \ldots, \alpha_m$, are different from zero, whereas the others vanish. Then we apply formula (5.5) to the exponential sum $\alpha_1 \cdot e^{\lambda_1 z} + \ldots + \alpha_m \cdot e^{\lambda_m z}$ and find

$$\frac{1}{2\pi} \int_0^{2\pi} \log |(\alpha x)| \ d\vartheta \sim \frac{L'}{2\pi} \cdot r,$$

L' being the length of the convex polygon $\mathfrak{D}'$ spanned around the points $P_1, \ldots, P_m$. Because they form part of our diagram it is geometrically clear that $L' \leqq L$; analytically this follows from the obvious inequality $h'(\vartheta) \leqq h(\vartheta)$ for the functions of support $h(\vartheta)$, $h'(\vartheta)$ of $\mathfrak{D}$ and $\mathfrak{D}'$ respectively. The defect under these circumstances is $\sim \frac{L-L'}{2\pi} \cdot r$.

The logarithm under the integral in (5.6) is half of the logarithm of the quotient $\frac{|\alpha|^2 \cdot |x|^2}{|(\alpha x)|^2}$. Both numerator and denominator are non-negative Hermitian forms of x, the numerator positive definite, the denominator at least semi-definite. Under the hypothesis $\alpha_1 \neq 0$ the diagonal coefficients of either form are positive. This suggests the following generalization of (5.6):

LEMMA 5. A. Let Δ_k be the opening $\vartheta_k' - \vartheta_k$ of the angle Θ_k if k is of first kind, $\Delta_k = 0$ otherwise, and denote by $\mathfrak{M}t_k$ the weighted average of any n numbers $t_1, \ldots, t_n$ as defined by

$$\mathfrak{M}t_k = \sum \Delta_k t_k / \sum \Delta_k = \frac{1}{2\pi} \sum \Delta_k t_k .$$

Let

$$G_0(x) = \sum g_{ij}^0 x_i \bar{x}_j, \qquad G(x) = \sum g_{ij} x_i \bar{x}_j$$

be two non-negative Hermitian forms whose principal coefficients g_{kk}^0, g_{kk} are actually positive. Then

$$(5.7) \qquad \frac{1}{2\pi} \int_0^{2\pi} \log G_0/G \cdot d\vartheta \longrightarrow \mathfrak{M}(\log g_{kk}^0/g_{kk}) \quad \text{for } r \to \infty.$$

We are primarily interested in the fact that the left side of (5.7) stays bounded; but with the same effort we can reach the sharper result of the lemma.*

---

*Cf. J. Weyl, Duke Math. Jour. 10, 1943, 123-143.

First an elementary geometric observation. An inequality

$$\Re(\lambda z) = ax + by \geqq 0 \qquad (\lambda = a-ib,\ z = x+iy)$$

defines a half plane $-\pi/2 \leqq \vartheta - \vartheta_0 \leqq \pi/2$. Shrink the half plane symmetrically to an angle of a little smaller opening than π, $|\vartheta - \vartheta_0| \leqq \pi/2 - \delta$. Then in that smaller angle

$$ax + by \geqq |\lambda| \cdot |z| \cdot \sin \delta.$$

Given a positive constant H we take off a (small) angle of opening H/r from either side of $\Theta_\mu^* = \Theta_k$ thus shrinking it to

$$\Theta_k(r) = \Theta_\mu^*(r): \qquad \vartheta_k + H/r \leqq \vartheta \leqq \vartheta_k' - H/r,$$

and we surround each of the dividing directions ϑ_μ^* by a symmetric angle of opening 2H/r,

$$\eta_\mu^*(r): \qquad -H/r \leqq \vartheta - \vartheta_\mu^* \leqq H/r.$$

Then the whole circumference of the circle of radius r is decomposed into alternating big arcs and small arcs $\Theta_\mu^*(r)$ and $\eta_\mu^*(r)$, provided r already exceeds the ν numbers $2H/\Delta_k$ ($\geqq 2H/\pi$) corresponding to the indices k of first kind, and the integral (5.7) is split into a major and a minor part, extending over the $\Theta_\mu^*(r)$ and the $\eta_\mu^*(r)$ respectively.

Let us examine the integral over one of the big arcs $\Theta_\mu^*(r) = \Theta_k(r)$ and let $\frac{\pi}{2}\beta$ be the least of the n - 1 moduli $|\lambda_k - \lambda_i|$, ($i \neq k$). By applying our elementary remark to the difference $(a_k - a_i)x + (b_k - b_i)y$ we find that

$$r(a_k \cos \vartheta + b_k \sin \vartheta) - r(a_i \cos \vartheta + b_i \sin \vartheta)$$
$$\geqq \frac{\pi}{2}\beta r \sin H/r \geqq \beta H$$

for ϑ in $\Theta_k(r)$ and $i \neq k$. Consequently $x_k\bar{x}_k$ exceeds all other products $x_i\bar{x}_j$ for $z = re^{i\vartheta}$, ϑ in $\Theta_k(r)$, to such a degree that

$$|x_i\bar{x}_j| \leqq e^{-\beta H} x_k\bar{x}_k$$

(except for the combination $i = k$, $j = k$). Given any (small) positive ϵ we may therefore ascertain H so big that on the arc $\Theta_k(r)$ of the circle K_r the quotient G_0/G differs from g^0_{kk}/g_{kk} by a factor between $1 + \epsilon$ and $(1+\epsilon)^{-1}$, and that for every one of the ν indices k of first kind. This done, the major part differs by less than ϵ from the sum

$$\frac{1}{2\pi}\sum(\Delta_k - 2H/r) \log g^0_{kk}/g_{kk} = \mathfrak{M}(\log g^0_{kk}/g_{kk}) - \frac{\text{Const.}}{r}$$

extending over the indices k of first kind. This holds irrespective of the value of r.

In the second place we are going to show that the integral of $\log G_0/G$ over any one of the small arcs $\eta^*_\mu(r)$ and thus the whole minor part lies between limits $\pm \frac{\text{Const.}}{r}$. The result then is that left and right members of (5.7) differ in absolute value by less than $\epsilon + \frac{\text{Const.}}{r}$ and thus by less than 2ϵ, as soon as r exceeds a certain value $r(\epsilon)$, quod erat demonstrandum.

Without loss of generality we may assume $\vartheta^*_\mu = 0$ and the first among the expressions $P_k(0)$ to have the maximum value $h(0)$:

$$a_1 = \dots = a_m = a, \qquad a_k < a \qquad \text{for } k = m + 1, \dots, n.$$

Because $a_1, \dots, a_m$ are equal, the numbers $b_1, \dots, b_m$ must be distinct. Moreover, since G_0/G is not altered by substituting $e^{-\lambda_1 z} \cdot x_k$ for x_k we may further assume $a = 0$, $b_1 = 0$, so that P_1 is the origin and

$$a_k = 0 \text{ for } k = 1, \dots, m; \quad a_k < 0 \text{ for } k = m + 1, \dots, n.$$

It suffices to study the integral

$$(5.8) \qquad \int_{-H/r}^{H/r} \log G \cdot d\vartheta.$$

Set $r\vartheta = t$, so that

$$(5.9) \qquad z = r \cdot e^{it/r}.$$

For t on the segment $\mathfrak{g}$, $-H \leqq t \leqq H$, we find, because of $a_k \leqq 0$ (and $|\vartheta| \leqq \pi/2$),

$$|x_k| = e^{a_k r \cos \vartheta + b_k r \sin \vartheta} \leqq e^{b_k r \sin \vartheta}$$
$$\leqq e^{|b_k t|} \leqq e^{|b_k| H} \qquad \text{for } k = 1, \dots, n.$$

This gives a simple constant upper bound for G along the arc $\eta_\mu^*(r)$ of the circle K_r, one that is independent of the radius r, and thus an upper bound of the form Const./r for the integral (5.8).

The delicate point is the lower bound. Because of the semi-definite character of G the form

$$G(x) - g_{11}|g_1x_1+\dots+g_nx_n|^2 \quad (g_1 = 1, \quad g_i = g_{i1}/g_{11}),$$

which no longer involves the variable x_1, is also semi-definite; hence

$$G(x) \geqq g_{11}|g_1x_1+\dots+g_nx_n|^2,$$

and our problem is reduced to finding a sufficiently sharp lower bound for

$$(5.10) \qquad \int_{-H/r}^{H/r} \log |g_1x_1+\dots+g_nx_n| \cdot d\vartheta.$$

Preparing to apply Lemma 4. C we now let t assume complex values but limit it to a bounded region $\mathfrak{G}$ surrounding the segment $\mathfrak{g}$. In that region we find for the modulus of

$$z - r = r(e^{it/r}-1) = i\int_0^t e^{i\tau/r} \cdot d\tau$$

an upper bound H' which is independent of t and r, and thus after the substitution (5.9)

$$(5.11) \qquad |x_k| \leqq e^{a_k r} \cdot e^{|\lambda_k| H'} = \text{Const.}\ e^{a_k r}.$$

For k = 1, ... , m the explicit expressions

$$x_k = e^{-ib_k z} = e^{-ib_k r} \cdot \exp\{-ib_k r(e^{it/r}-1)\},$$

in particular $x_1 = 1$, will be used. Set

$$f_1(r;t) = 1 + (g_{m+1}x_{m+1}+\dots+g_n x_n), \quad f_1(t) = 1;$$

$$f_k(r;t) = \exp\{-ib_k r(e^{it/r}-1)\},\ f_k(t) = \exp\ (b_k t) \qquad (k=2,\dots,m).$$

$f_k(r;t)$ tends to $f_k(t)$ (for k=2,...,m) with $r \longrightarrow \infty$ uniformly for $t \in \mathfrak{G}$. In the same manner $f_1(r;t)$ tends to $f_1(t)$ because of (5.11) and $a_k < 0$ for k = m + 1, ... , n. Our integral (5.10) assumes the form

$$(5.12) \quad \frac{1}{r} \int_{-H}^{H} \log |\xi_1(r)\cdot f_1(r;t)+\dots+\xi_m(r)\cdot f_m(r;t)|\ dt$$

where the coefficients are "rotating" functions of the parameter r,

$$(5.13) \qquad \xi_k(r) = g_k \cdot e^{-ib_k r} \qquad (k=1,\dots,m)$$

(in particular, $\xi_1(r) = 1$). We apply Lemma 4. C to the integral

$$(5.14) \quad \int_{-H}^{H} \log |\xi_1 f_1(r;t)+\dots+\xi_m f_m(r;t)|\ dt$$

in which the parameters $\xi_1, \dots, \xi_m$ vary freely over the torus

$$(5.15) \qquad |\xi_k| = |g_k| = l_k \qquad (k=1,\dots,m).$$

Notice that $l_1 = 1$ and that the m functions $f_k(t) = \exp(b_k t)$ are linearly independent. The lemma yields a lower bound -B for (5.14) valid for sufficiently large r and all $(\xi_1, \ldots, \xi_m)$ on the torus, and since the point (5.13) never leaves the torus, -B/r is a lower bound for (5.12). This is the decisive step in our argument: let the point $(\xi_1, \ldots, \xi_m)$ roam freely over the torus without binding it to the complicated Lissajous figure described by the point (5.13) on the torus.

From the order T we pass to the order T_p of rank p. The components of $X^p = [x, x', \ldots, x^{(p-1)}]$ are

$$X^p(i_1, \ldots, i_p) = \begin{vmatrix} 1, & \lambda_{i_1}, & \ldots, & \lambda_{i_1}^{p-1} \\ \cdot & \cdot & \cdot & \cdot \\ 1, & \lambda_{i_p}, & \ldots, & \lambda_{i_p}^{p-1} \end{vmatrix} \exp\{(\lambda_{i_1} + \ldots + \lambda_{i_p})z\},$$

$$(i_1 < \ldots < i_p).$$

The constant factor in front of the exponential function does not vanish. Hence $\mathfrak{C}_p$ is an exponential curve of essentially the same type as $\mathfrak{C}$, and we find

$$T_p(r) \sim \frac{L_p}{2\pi} \cdot r \tag{5.16}$$

where L_p is the length of the convex polygon spanned around the $\binom{n}{p}$ points

$$(a_{i_1} + \ldots + a_{i_p}, \; b_{i_1} + \ldots + b_{i_p}) \qquad (i_1 < \ldots < i_p) \tag{5.17}$$

To be sure these points are not necessarily distinct, but the argument leading to the formula (5.3) did not presuppose distinctness of the exponents λ. The function of support $h_p(\vartheta)$ of the points (5.17) is equal to the largest of the $\binom{n}{p}$ values

$$P_{i_1}(\vartheta) + \ldots + P_{i_p}(\vartheta)$$

for any ϑ. Arrange the n numbers

$$P_1(\vartheta), \ldots, P_n(\vartheta)$$

in decreasing order and denote them in this order by

$$q_1(\vartheta) \geqq q_2(\vartheta) \geqq \ldots \geqq q_n(\vartheta).$$

Then

$$h_p(\vartheta) = q_1(\vartheta) + \ldots + q_p(\vartheta)$$

and we find

$$L_p = \int_0^{2\pi} \{q_1(\vartheta)+\ldots+q_p(\vartheta)\}\, d\vartheta. \tag{5.18}$$

Since the Wronskian of the exponential curve vanishes nowhere, the curve is without stationary points of any rank.

Instead of (5.1) we could discuss a more general exponential curve $\mathfrak{C}$ with coordinates $y_1, \ldots, y_n$, defined by equations

$$y_i = a_{i1}e^{\lambda_1 z} + \ldots + a_{iN}e^{\lambda_N z}. \tag{5.19}$$

We naturally assume that the exponents $\lambda_1, \ldots, \lambda_N$ are distinct, and that no single column of the matrix

$$\begin{vmatrix} a_{11}, & \ldots, & a_{1N} \\ \cdot & \cdot\ \cdot\ \cdot & \cdot \\ a_{n1}, & \ldots, & a_{nN} \end{vmatrix}$$

vanishes. We may consider this curve $\mathfrak{C}$ as a projection of the curve $\mathfrak{C}^{(N)}$ in N-space,

$$x_1 = e^{\lambda_1 z}, \ldots, x_N = e^{\lambda_N z}$$

upon an n-dimensional subspace. By the substitution

$$y_i = a_{i1}x_1 + \dots + a_{iN}x_N$$

the square sum $|y_1|^2 + \dots + |y_n|^2$ changes into a semi-definite Hermitian form G of the variables $x_1, \dots, x_N$ the principal coefficients of which are actually positive. Applying our lemma to this G and $G_0 = |x_1|^2 + \dots + |x_N|^2$ we realize at once that $\mathfrak{C}$ and $\mathfrak{C}^{(N)}$ are of the same order, and thus the order function T(r) of $\mathfrak{C}$ is equivalent to $\frac{L}{2\pi}\cdot r$ where L is the length of the convex polygon surrounding the points $\bar{\lambda}_1, \dots, \bar{\lambda}_N$. For any plane $\beta_1 y_1 + \dots + \beta_n y_n = 0$ we find

$$N(r;\beta) \sim \frac{L'}{2\pi}\cdot r,$$

L' denoting the length of the convex polygon spanned around those among the points $\bar{\lambda}_K$ for which

$$\beta_1 a_{1K} + \dots + \beta_n a_{nK} \neq 0,$$

and hence the corresponding defect

$$m(r;\beta) \sim \frac{L-L'}{2\pi}\cdot r.$$

In a number of ways the example of exponential curves, even of the general exponential curves of type (5.19), proves to be more easily manageable and more informative than that of the rational curves.

§6. Kronecker multiplication. Intersections with an algebraic surface

Suppose we have a meromorphic curve $\mathfrak{C}_1$ with the homogeneous coordinates $x_1^{(1)}, \dots, x_n^{(1)}$ in n-space and a curve $\mathfrak{C}_2$ with the coordinates $x_1^{(2)}, \dots, x_m^{(2)}$ in m-space. Let the branches associated with the point $z = z_0$ be given in reduced representations,

$$(6.1) \qquad x_i^{(1)} = \mathfrak{P}_i^{(1)}(z-z_0), \qquad x_k^{(2)} = \mathfrak{P}_k^{(2)}(z-z_0).$$

We form the n·m coordinates $x_{i,k} = x_i^{(1)} x_k^{(2)}$ with the corresponding expansions

$$(6.2) \qquad x_{i,k} = \mathfrak{P}_i^{(1)} \mathfrak{P}_k^{(2)}(z-z_0).$$

They constitute a reduced representation, and multiplication of the two representations (6.1) by arbitrary gauge factors $\rho^{(1)}$, $\rho^{(2)}$ has the effect of multiplying (6.2) by the gauge factor $\rho^{(1)}\rho^{(2)}$. Hence (6.2) defines a meromorphic curve in n·m dimensions which is called the Kronecker product $\mathfrak{C}_1 \times \mathfrak{C}_2$.

THEOREM. The order $T\{\mathfrak{C}\}$ of the Kronecker product $\mathfrak{C} = \mathfrak{C}_1 \times \mathfrak{C}_2$ equals $T\{\mathfrak{C}_1\} + T\{\mathfrak{C}_2\}$.

In other words, the order shows a logarithmic behavior under multiplication of curves. The proof is very simple. The order of $\mathfrak{C}$ could be computed by an arbitrary non-vanishing linear form

$$\sum_{i,k} \alpha_{ik} x_{i,k} = \sum_{i,k} \alpha_{ik} x_i^{(1)} x_k^{(2)}.$$

But we specialize the coefficients α_{ik} as follows: $\alpha_{ik} = \alpha_i^{(1)} \alpha_k^{(2)}$; then

$$(6.3) \qquad \sum \alpha_{ik} x_{i,k} = \sum \alpha_i^{(1)} x_i^{(1)} \sum \alpha_k^{(2)} x_k^{(2)}.$$

Thus the order in which the left member vanishes for $z = z_0$ is the sum of the orders of the two factors at the right-hand side, and moreover by combining (6.3) with

$$\sum |\alpha_{ik}|^2 = \sum |\alpha_i^{(1)}|^2 \cdot \sum |\alpha_k^{(2)}|^2, \quad \sum |x_{i,k}|^2 = \sum |x_i^{(1)}|^2 \cdot \sum |x_k^{(2)}|^2$$

we find

$$\log \frac{1}{\|\alpha x\|} = \log \frac{1}{\|\alpha^{(1)}x^{(1)}\|} + \log \frac{1}{\|\alpha^{(2)}x^{(2)}\|} .$$

Therefore

$$N(R;\alpha) = N^{(1)}(R;\alpha^{(1)}) + N^{(2)}(R;\alpha^{(2)}),$$
$$m(r;\alpha) = m^{(1)}(r;\alpha^{(1)}) + m^{(2)}(r;\alpha^{(2)}).$$

Multiplication by itself of the curve $\mathfrak{C}$ in n-space with the coordinates x_i gives rise to a curve with the n^2 coordinates $x_i x_k$. However, in this special case it is not reasonable to carry $x_i x_k$ and $x_k x_i$ (for $i \neq k$) as two different coordinates. Hence for the fth power $\mathfrak{C}^f$ of $\mathfrak{C}$, we use the monomials

$$x(f_1,\ldots,f_n) = \sqrt{\left(\frac{f!}{f_1!\ldots f_n!}\right)} \cdot x_1^{f_1} \ldots x_n^{f_n} \tag{6.4}$$
$$(f_1 \geqq 0,\ldots,f_n \geqq 0, \quad f_1 + \ldots + f_n = f)$$

as coordinates. The numerical factors are added to bring about the relation

$$\sum x(f_1,\ldots,f_n) \cdot \overline{x}(f_1,\ldots,f_n) = (x_1\overline{x}_1 + \ldots + x_n\overline{x}_n)^f .$$

We then have the simple statement:

The order of $\mathfrak{C}^f$ is f-times the order of $\mathfrak{C}$.

A linear form of the coordinates (6.4),

$$\sum \alpha(f_1,\ldots,f_n) \cdot x(f_1,\ldots,f_n) \tag{6.5}$$

is a homogeneous form $F(x_1,\ldots,x_n)$ of degree f of the variables $x_1,\ldots,x_n$. The specialization by which we prove our theorem is indicated by the formulas

$$\alpha(f_1,\ldots,f_n) = \sqrt{\left(\frac{f!}{f_1!\ldots f_n!}\right)} \cdot \alpha_1^{f_1} \ldots \alpha_n^{f_n} \quad \text{or}$$
$$\sum \alpha(f_1,\ldots,f_n) \cdot x(f_1,\ldots,f_n) = (\alpha_1 x_1 + \ldots + \alpha_n x_n)^f .$$

The intersections of $\mathfrak{C}^f$ with the plane defined by the vanishing of any linear form (6.5) are nothing else but the intersections of $\mathfrak{C}$ with the algebraic surface of order f, $F(x_1,\ldots,x_n) = 0$. Hence our result is a generalization of the fact that an algebraic curve of order ν intersects an algebraic surface of order f in $f \cdot \nu$ points.

§7. Projection

In general the central projection of an algebraic curve from a point has the same order ν as the original curve. However, if the center of projection lies on the curve the order is reduced by the multiplicity of the intersection between point and curve. This theorem may be generalized in two ways, - by substituting any h-element as center of projection for a point or 1-element, and by studying the order ν_p or rank p instead of ν. It is in this generality that we are going to formulate the corresponding fact for meromorphic curves $\mathfrak{C}$.

Let $e_1,\ldots,e_n$ be a normal coordinate system. Projection from the unit h-spread $\{E^h\} = \{e_1,\ldots,e_h\}$ changes a vector $(x_1,\ldots,x_n)$ into $(x_{h+1},\ldots,x_n)$ and any (special) p-ad X^p into $\tilde{X}^p$ where $\tilde{X}^p(i_1,\ldots,i_p) = X(i_1,\ldots,i_p)$ with the range $h+1, \ldots, n$ for the indices i. Set $E^h = [e_1,\ldots,e_h]$ so that $|E^h| = 1$. We compute the order $T_p(R)$ of rank p of the given meromorphic curve $\mathfrak{C}$ by means of a linear form (AX^p) in which the contravariant $A = [\alpha_1\ldots\alpha_p]$ is spanned by vectors α_q of the special form $\alpha = (0,\ldots,0,\alpha_{h+1},\ldots,\alpha_n)$, so that $A(i_1,\ldots,i_p)$ vanishes if one of the figures $1,\ldots,h$ appears among the indices $i_1,\ldots,i_p$. Normalizing by $|A| = 1$ we have

$$N\left(R, \frac{|(AX^p)|}{|X^p|}\right) + \left[\frac{1}{2\pi}\int_0^{2\pi} \log \frac{|X^p|}{|(AX^p)|}\, d\vartheta\right]_{r_0}^{R} = T_p(R).$$

The corresponding order $\tilde{T}_p(R) = T_p(R;E^h)$ of the projected curve $\mathfrak{C}$ is the same expression, except that $|X^p|$ is to be replaced by $|\tilde{X}^p|$. Observing that $|\tilde{X}^p| = |[E^h, X^p]|$ we thus find

$$T_p(R) - T_p(R;E^h) = N\left(R, \frac{|[E^h X^p]|}{|X^p|}\right) + \left[\frac{1}{2\pi}\int_0^{2\pi} \log \frac{|X^p|}{|[E^h X^p]|} \cdot d\vartheta\right]_{r_0}^{R}.$$

The auxiliary contravariant p-ad A has disappeared. In this form the equation holds for any h-element $\{E^h\}$ with the normalization $|E^h| = 1$ because proceeding from a given unitary metric we can always construct a normal coordinate system $e_1,\dots,e_n$ such that its first h vectors $e_1,\dots,e_h$ span the given $\{E^h\}$. The first term in the right member, which we denote by $\tilde{N}_p(R;E^h)$, may be written as

$$(7.1) \qquad \tilde{N}_p(R;E^h) = \sum\nolimits_z \nu_p(z:E^h) \cdot \phi(R;z)$$

if $d_p(z_0) + \nu_p(z_0:E^h)$ is the order in which the components of $[E^h X^p]$ vanish simultaneously at $z = z_0$. It is thus the valence of the intersections of the center of projection $\{E^h\}$ with $\mathfrak{C}_p$ in the circle of radius R, whereas the the non-negative

$$(7.2) \qquad \tilde{m}_p(r;E^h) = \frac{1}{2\pi}\int_0^{2\pi} \log \frac{1}{\|E^h : X^p\|} \cdot d\vartheta$$

plays the role of the corresponding compensating term. The latter depends on the choice of the unitary metric. However, by passing to another unitary metric it is changed by an additive function of r varying between the fixed limits $\pm \log \lambda_{\min(p,h)}$, as computed in Chapter I, §5. Thus we have arrived at the following formula describing the relation of the orders of rank p of a given curve and its projection from $\{E^h\}$:

$$(7.3) \quad T_p(R) - T_p(R;E^h) = \tilde{N}_p(R;E^h) + [\tilde{m}_p(r;E^h)]_{r_0}^{R}.$$

We were somewhat arbitrary to normalize an additive constant in the order function by the condition $T(r_0) = 0$. If we deviate from this for the projected curve and introduce $T_p(R;E^h) - \tilde{m}_p(r_0;E^h)$ as its order of rank p, then it is true that the order of the projected curve does not exceed that of the original curve, and the "loss by projection"

$$T_p(R) - \{T_p(R;E^h) - \tilde{m}_p(r_0;E^h)\} = \Lambda_p(R;E^h)$$

is given by

$$\Lambda_p(R;E^h) = \tilde{N}_p(R;E^h) + \tilde{m}_p(R;E^h).$$

The quantities (7.1), (7.2) coincide with (2.6), (2.6') in the highest case $h = n - p$ if $\{A\}$ is the dual of $\{E^{n-p}\}$:

$$N_p(R;A^p) = \tilde{N}_p(R;A^{n-p}), \quad m_p(r;A^p) = \tilde{m}_p(r;A^{n-p})$$

$$\text{for } A^p = *A^{n-p}.$$

Suppose $\{E^{h-1}\} \subset \{E^h\}$. The multiplicity $\nu_p(z_0:E^h)$ with which $[E^h, X^p_{red}]$ vanishes at z_0 cannot be less than the vanishing order $\nu_p(z_0:E^{h-1})$ of $[E^{h-1}, X^p_{red}]$. Hence $\nu_p(z_0:E^h) \geqq \nu_p(z_0:E^{h-1})$,

$$\tilde{N}_p(R;E^h) \geqq \tilde{N}_p(R;E^{h-1}).$$

The relation (I, 4.1) or

$$\log \frac{1}{\| E^h : X^p \|} \geqq \log \frac{1}{\| E^{h-1} : X^p \|}$$

yields

$$\tilde{m}_p(r;E^h) \geqq \tilde{m}_p(r;E^{h-1}).$$

By addition

$$\Lambda_p(R;E^h) \geqq \Lambda_p(R;E^{h-1}).$$

More surprising than these facts are the consequences of the inequality

$$\|E^h:X^p\| \leq \|E^h:X^{p-1}\| \tag{7.4}$$

which follows by the same lemma I, 4.A from $\{X^{p-1}\} \subset \{X^p\}$. It implies that the order $\nu_p(z_0:E^h)$ with which $\|E^h:X^p\|$ vanishes cannot be less than $\nu_{p-1}(z_0:E^h)$. The obvious remark that $[E^hX^p]$ vanishes at least in the same order as $[E^hX^{p-1}]$ yields only the inequality

$$\{d_p(z_0) - d_{p-1}(z_0)\} + \{\nu_p(z_0:E^h) - \nu_{p-1}(z_0:E^h)\} \geqq 0.$$

Our stronger result leads to

$$\tilde{N}_p(R;E^h) \geqq \tilde{N}_{p-1}(R;E^h), \tag{7.5}$$

and moreover, by taking the logarithm and integrating, (7.4) gives rise to the complementary inequality

$$\tilde{m}_p(r;E^h) \geqq \tilde{m}_{p-1}(r;E^h). \tag{7.6}$$

In consequence of the last two relations, <u>the loss</u> $\Lambda_p(R;E^h)$ <u>by projection along</u> $\{E^h\}$ <u>of rank</u> p <u>is not smaller than the same loss</u> $\Lambda_{p-1}(R;E^h)$ <u>of rank</u> p - 1. These are quite remarkable facts.

§8. Poincaré's theorem for meromorphic functions

We do not pretend that the theory of meromorphic curves as yet comprises all important known facts about meromorphic functions. Indeed, of the Poincaré-Hadamard theory it includes only Hadamard's first theorem.

For this reason the two next sections dealing with Poincaré's and with Hadamard's second theorem form a sort of enclave in our pattern. We generalize these two propositions from entire to meromorphic functions and prove them in this generality, closely following R. Nevanlinna's approach in his fundamental paper 1925.

THEOREM. Suppose we are given a non-negative integer k and a number λ in the interval $k < \lambda \leqq k + 1$; moreover two sequences a_m, b_n of complex numbers $\neq 0$ such that $\sum |a_m|^{-\lambda}$, $\sum |b_n|^{-\lambda}$ converge. With an arbitrary polynomial P(z) of formal degree k form the function

$$f(z) = e^{P(z)} \cdot \frac{\prod_m E(k; z/a_m)}{\prod_n E(k; z/b_n)} \tag{8.1}$$

composed of Weierstrass's primary factors of type k,

$$E(k;u) = (1-u) \cdot \exp\left(\frac{u}{1} + \ldots + \frac{u^k}{k}\right). \tag{8.2}$$

Then the order T(R) of the meromorphic function f(z) satisfies the limit equation

$$T(R) = o(R^\lambda) \tag{8.3}$$

(and

$$\int^{\infty} T(r) \cdot r^{-(\lambda+1)} dr \tag{8.4}$$

is convergent provided λ is not integral).

Proof. We know that $T\{1/f\} = T\{f\}$. Moreover

$$T\{fg\} \leqq T\{f\} + T\{g\} + \text{const.}$$

for any meromorphic functions f, g. This follows from a combination of Kronecker multiplication with projection. One forms the product of the two curves in 2-space with the coordinates (x_1, x_2), (y_1, y_2) respectively; $f = x_1/x_2$, $g = y_1/y_2$. The product is a curve in 4-space with the coordinates x_1y_1, x_1y_2, x_2y_1, x_2y_2, but we use only its projection in 2-space with the two coordinates x_1y_1, x_2y_2. For the factor $e^{P(z)}$, P(z) = polynomial of formal degree k, we have $\log M(r) \leqq C \cdot r^k$ and thus

$$T(R) \leqq C \cdot R^k, \quad T(R) = o(R^\lambda).$$

For these reasons it is sufficient to prove the estimate

$$\log M(r) = o(r^\lambda)$$

for the canonical product

$$(8.5) \qquad f(z) = \prod_n E(k; z/a_n).$$

First we must try to estimate fairly accurately the primary factor $E(k;u)$. For small u, $\log |E(k;u)|$ is in first approximation $\frac{1}{k+1} |u|^{k+1}$, for large u it behaves like $\frac{1}{k} |u|^k$. Hence it is reasonable to expect something like the following

LEMMA 8. A.

$$\log |E(k;u)| \leqq B \frac{|u|^{k+1}}{1+|u|} \text{ if } k > 0.$$

The constant B may be chosen as

$$(8.6) \qquad B = 2(3+\log k).$$

[For $k = 0$ one has

$$\log |E(0;u)| \leqq \log (1+|u|).]$$

Take this lemma for granted and assume $k > 0$. Then the inequality

$$(8.7) \qquad \log M(r) \leqq B \cdot \sum \frac{r^{k+1}}{|a_n|^k (r+|a_n|)}$$

holds for the entire function (8.5). We suppose the $|a_n|$ to be arranged in ascending order. With the notation

$$|a|/r = v, \qquad \alpha = \lambda - k \leqq 1$$

we may write the general summand in (8.7) as

$$\frac{r^\lambda}{|a|^\lambda} \frac{v^\alpha}{1+v}.$$

Observe that the factor $v^{\alpha}/(1+v) \leqq 1$. Indeed $v^{\alpha} \leqq 1$, $1 + v \geqq 1$ for $0 \leqq v \leqq 1$, whereas $v^{\alpha} \leqq v$, $1 + v \geqq v$ for $v \geqq 1$. Consequently

$$\text{(8.8)} \qquad \log M(r) \leqq Br^{\lambda} \cdot \sum |a_n|^{-\lambda} = O(r^{\lambda}).$$

The improvement to $o(r^{\lambda})$ is obtained by splitting the sum in (8.7) into $\sum_1^N$ and $\sum_{N+1}^{\infty}$. For a fixed N the first part is obviously $O(r^k) = o(r^{\lambda})$ while for the second part we find by the same procedure followed above the upper bound

$$Br^{\lambda} \cdot \sum_{N+1}^{\infty} |a_n|^{-\lambda}.$$

For an arbitrarily given $\epsilon > 0$ we may fix N so that this becomes $< \epsilon \cdot r^{\lambda}$. After having done so the first part will also, from a certain r on, be less than $\epsilon \cdot r^{\lambda}$.

The case $k = 0$ is no exception. Indeed, if $0 < \lambda \leqq 1$ the inequality

$$\frac{x^{1-\lambda}}{1+x} \leqq 1 \text{ or } \frac{1}{1+x} \leqq x^{\lambda-1} \qquad \text{for } x \geqq 0$$

implies by integration

$$\log (1+x) \leqq \frac{1}{\lambda} x^{\lambda}.$$

Hence

$$\text{(8.9)} \quad \log M(r) \leqq \sum \log \left(1 + \frac{r}{|a_n|}\right) \leqq \frac{1}{\lambda} r^{\lambda} \cdot \sum |a_n|^{-\lambda}$$

in complete analogy to (8.8).

In carrying out the integration required by the supplementary statement (8.4) of our theorem we make use of the values of the following integrals:

$$\int_0^{\infty} \frac{u^{k-\lambda}}{1+u}\, du = \frac{\pi}{\sin \pi(k+1-\lambda)} = \frac{\pi}{|\sin \pi\lambda|} \quad \text{for } k < \lambda < k+1,$$

$$\int_0^{\infty} u^{-\lambda-1} \log (1+u)du = \frac{1}{\lambda} \int_0^{\infty} \frac{u^{-\lambda} du}{1+u} = \frac{1}{\lambda} \cdot \frac{\pi}{\sin \pi\lambda} \text{ for } 0<\lambda< 1.$$

Hence (8.7) for $k > 0$ and the corresponding inequality (8.9) for $k = 0$ yield

$$\int_{r_0}^{R} r^{-(\lambda+1)} \log M(r) \cdot dr \leqq \frac{B\pi}{|\sin \pi\lambda|} \sum |a_n|^{-\lambda} \qquad (k>0,\ k<\lambda<k+1)$$

$$\text{or} \leqq \frac{\pi}{\lambda \sin \pi\lambda} \sum |a_n|^{-\lambda} \qquad (0<\lambda<1)$$

respectively.

It remains to prove the elementary lemma. For $|z| < 1$ we may write

$$\log E(k;z) = \log (1-z) + \frac{z}{1} + \ldots + \frac{z^k}{k} = -\int_0^z \frac{t^k dt}{1-t} ,$$

hence for $|z| \leqq a < 1$,

$$\log |E(k;z)| \leqq |\log E(k;z)| \leqq \frac{1}{1-|z|} \cdot \frac{|z|^{k+1}}{k+1}$$

(8.10)
$$\leqq \frac{1+a}{1-a} \cdot \frac{1}{k+1} \cdot \frac{|z|^{k+1}}{1+|z|}$$

On the other hand for all z

$$|E(k;z)| \leqq (1+|z|) \cdot \exp \left(\frac{|z|}{1} + \ldots + \frac{|z|^k}{k}\right)$$

$$\leqq \exp \left(2 \cdot \frac{|z|}{1} + \frac{1}{2}|z|^2 + \ldots + \frac{1}{k}|z|^k\right)$$

and thus

$$\log |E(k;z)| \leqq |z|^k \left(\frac{1}{k} + \frac{1}{k-1} \cdot \frac{1}{a} + \ldots + \frac{1}{2} \cdot \frac{1}{a^{k-2}} + 2 \cdot \frac{1}{a^{k-1}}\right)$$

for $|z| \geqq a$. The right member equals

$$\frac{|z|^k}{a^{k-1}} \left(2 + \frac{a}{2} + \ldots + \frac{a^{k-1}}{k}\right) \leqq \frac{|z|^k}{a^{k-1}} \left(2 + \frac{1}{2} + \ldots + \frac{1}{k}\right)$$

(8.11)
$$\leqq \frac{|z|^k}{a^{k-1}} (2 + \log k) \leqq \frac{|z|^{k+1}}{1+|z|} \cdot \frac{1+a}{a^k} (2 + \log k).$$

Any positive a less than unity will do for proving the lemma. A fairly good choice which nearly equalizes the two factors in (8.10) and (8.11), namely

$$(8.12) \qquad \frac{1}{k+1}\cdot\frac{1+a}{1-a} \qquad \text{and} \qquad (2 + \log k)\cdot\frac{1+a}{a^k},$$

is

$$a = (3k + k \log k)/(1 + 3k + k \log k)$$

which makes

$$\frac{1}{1-a} = 1 + 3k + k \log k \leqq (k + 1)(3 + \log k),$$

while application of the inequalities

$$(1 + \frac{x}{k})^k < (1 - \frac{x}{k})^{-k} \leqq (1 - x)^{-1} \qquad (0<x<1)$$

to $x = 1/(3 + \log k)$ results in the relation

$$(\frac{1}{a})^k < \frac{3 + \log k}{2 + \log k}.$$

Moreover $1 + a < 2$. Thus both factors (8.12) become less than (8.6).

§9. Hadamard's second theorem for meromorphic functions

THEOREM. A meromorphic function f(z) with the zeros $a_m \neq 0$ and poles $b_n \neq 0$ for which $T(R) = o(R^{k+1})$ is necessarily of the form

$$(9.1) \qquad f(z) = e^{P(z)}\cdot\lim_{R\to\infty}\frac{\prod_{|a_m|\leqq R} E(k;z/a_m)}{\prod_{|b_n|\leqq R} E(k;z/b_n)}$$

where P(z) is a polynomial of formal degree k and convergence is uniform in any fixed circle $|z| \leqq r$.

We have seen (in §3) that the quantities $N_0(R)$ and $N_\infty(R)$ referring to the zeros and poles of f stay below

T(R). Hence the convergence of the integral

$$\int^{\infty} r^{-(k+2)} \cdot T(r)\, dr$$

would imply the convergence of the series

$$\sum |a_m|^{-k-1}, \qquad \sum |b_n|^{-k-1}$$

and thus of the Weierstrass canonical products

$$\prod_m E(k; z/a_m) \quad \text{and} \quad \prod_n E(k; z/b_n).$$

The slightly different hypothesis of our theorem makes it necessary to combine numerator and denominator as in (9.1) before passing to the limit.

In order to prove (9.1) we first derive Jensen's formula from Poisson's integral, and then follow R. Nevanlinna in transforming Jensen's formula into (9.1). Given any analytic function f(z) in the closed unit circle $|z| \leqq 1$, Poisson's integral shows how to express f(z) in the interior of the circle in terms of the boundary values of its real part:

$$(9.2) \qquad f(z) - ic = \frac{1}{2\pi} \int_0^{2\pi} \frac{\zeta + z}{\zeta - z} \cdot \Re f(\zeta) \cdot d\vartheta$$

$$(|z| < 1, \ \zeta = e^{i\vartheta}).$$

c denotes the imaginary part of f(0). Replace the unit circle by the circle $|z| \leqq R$ of radius R and apply (9.2) to log f(z) under the assumption that the analytic function f(z) is without zeros and poles and hence log f(z) single-valued in that circle:

$$(9.3) \quad \log f(z) - ic = \frac{1}{2\pi} \int_0^{2\pi} \frac{\zeta + z}{\zeta - z} \cdot \log |f(\zeta)| \cdot d\vartheta$$

$$(\zeta = R \cdot e^{i\vartheta}, \ |z| < R).$$

Next consider any meromorphic function with the zeros a_m and poles b_n. For any point a within the circle K_R of radius R the linear function

$$\frac{R(z-a)}{R^2-\bar{a}z}$$

has the zero a and is of modulus 1 along the periphery. We apply (9.3) to the quotient of f(z) and

$$\prod_{|a|_m<R} \frac{R(z-a_m)}{R^2-\bar{a}_m z} \Big/ \prod_{|b_n|<R} \frac{R(z-b_n)}{R^2-\bar{b}_n z}$$

which is without zeros and poles in K_R and thus obtain Jensen's formula

$$\log f(z) - ic = \sum_{|a_m|<R} \log \frac{R(z-a_m)}{R^2-\bar{a}_m z} - \sum_{|b_n|<R} \log \frac{R(z-b_n)}{R^2-\bar{b}_n z}$$

$$+ \frac{1}{2\pi} \int_0^{2\pi} \frac{\zeta+z}{\zeta-z} \log |f(\zeta)| \cdot d\vartheta$$

$$(\zeta = Re^{i\vartheta},\ |z| < R).$$

Just as in the case of the condenser formula, the result holds good even for those "critical" circles whose periphery $|z| = R$ is not free from zeros or poles. But if one does not want to be bothered by this possibility, one might exclude the critical values of R in the ensuing argument.

We form the $(k+1)^{th}$ derivative $D^{(k+1)}$ of Jensen's formula and find

$$D^{(k+1)}\log f(z) =$$

$$(-1)^k k! \left\{ \sum_{|a_m|<R} (z-a_m)^{-(k+1)} - \sum_{|b_n|<R} (z-b_n)^{-(k+1)} - \mu_R(z) \right\}$$

where

$$\mu_R(z) = \sum_{|a_m|<R} (z - \frac{R^2}{\bar{a}_m})^{-(k+1)} - \sum_{|b_n|<R} (z - \frac{R^2}{\bar{b}_n})^{-(k+1)}$$

$$- \frac{k+1}{2\pi} \int_0^{2\pi} \frac{2\zeta}{(\zeta - z)^{k+2}} \log |f(\zeta)| \cdot d\vartheta.$$

(Before differentiating, write

$$\frac{\zeta + z}{\zeta - z} = \frac{2\zeta}{\zeta - z} - 1.)$$

I maintain that under the hypothesis of our theorem

$$(9.4) \qquad \lim_{R \to \infty} \mu_R(z) = 0$$

uniformly with respect to z if z varies in any circle of fixed radius r.

Taking this for granted, we reverse the differentiation $D^{(k+1)}$ by integrating $(k+1)$-times from 0 to z. In this way $D^{(k+1)}\log f(z)$ gives rise to $\log f(z) - P(z)$ where $P(z)$ is a certain polynomial of formal degree k, namely the one consisting of the first $k + 1$ terms of the Taylor series of $\log f(z)$. The term $(z-a)^{-(k+1)}$ changes into

$$\log (1 - \frac{z}{a}) + \{\frac{1}{1}\frac{z}{a} + \ldots + \frac{1}{k} (\frac{z}{a})^k\} = \log E(k;\frac{z}{a}).$$

Hence we find that

$$P(z) + \log \frac{\prod_{|a_m|<R} E(k;z/a_m)}{\prod_{|b_n|<R} E(k;z/b_n)} \longrightarrow \log f(z)$$

with $R \longrightarrow \infty$ uniformly for $|z| \leq r$.

The proof of our theorem is thus reduced to the limit equation (9.4). First compare the sum

$$(9.5) \qquad \sum_{|a_m|<R} \left|z - \frac{R^2}{\bar{a}_m}\right|^{-(k+1)}$$

with

$$\sum_{|a_m|<R} 1 = n_0(R).$$

Since $|R^2/\bar{a}_m| > R$ and $|z| \leqq r$, (9.5) is less than

$$\frac{n_0(R)}{(R-r)^{k+1}}.$$

Repeat for $\lambda = k + 1$ the argument used to prove Lemma 3. A:

$$\frac{n_0(R)}{(k+1)R^{k+1}}\left(1 - \frac{1}{2^{k+1}}\right) \leqq \int_R^{2R} \frac{n_0(r)dr}{r^{k+2}} = \int_R^{2R} \frac{dN_0(r)}{r^{k+1}}$$

$$\leqq \frac{N_0(2R) - N_0(R)}{R^{k+1}} \leqq \frac{N_0(2R)}{(2R)^{k+1}} \cdot 2^{k+1}.$$

But $N_0(R) \leqq T(R)$ + const., and hence

$$T(R)/R^{k+1} \longrightarrow 0$$

implies

$$n_0(R)/R^{k+1} \longrightarrow 0 \text{ for } R \longrightarrow \infty.$$

It remains to discuss the integral in $\mu_R(z)$. Observe that

$$|\log |f|| = \log^+|f| + \log^+|1/f|.$$

Hence the absolute value of that integral cannot exceed

$$(9.6) \qquad \frac{2R}{(R-r)^{k+2}} \left\{\int_0^{2\pi} \log^+|f(\zeta)| \cdot d\vartheta + \int_0^{2\pi} \log^+|1/f(\zeta)| \cdot d\vartheta\right\}.$$

$$(\zeta = Re^{i\vartheta}).$$

Because of

$$\log^+|f| \leqq \log \sqrt{1+|f|^2}, \ \log^+(1/|f|) \leqq \log \sqrt{1+|1/f|^2}$$

and equations (3.3), both integrals in (9.6) are less than T(R) + const.

CHAPTER III

THE SECOND MAIN THEOREM FOR MEROMORPHIC CURVES

§1. The formula of the second main theorem

We return to the general theory of meromorphic curves and propose to develop the analogue of Plücker's formulas.

Let us form the second differences of the equations (II, 2.10). Because

$$d_{p+1}(z_0) - 2d_p(z_0) + d_{p-1}(z_0) = v_p(z_0) - 1,$$

the valence in the circle of radius R of the stationary points of rank p,

$$(1.1) \qquad V_p(R) = \sum_z (v_p(z)-1)\phi(R;z)$$

enters into the resulting equation

$$(1.2) \quad V_p(R) + \{T_{p+1}(R)-2T_p(R)+T_{p-1}(R)\} = [\Omega_p(r)]_{r_0}^{R}$$

in which

$$(1.3) \qquad \Omega_p(r) = \frac{1}{2\pi}\int_0^{2\pi} \log \frac{|X^{p+1}|\cdot|X^{p-1}|}{|X^p|^2}\cdot d\vartheta.$$

Like all valences $V_p(R)$ is a function of regular type. We have derived this formula from the normalized global representation of the meromorphic curve, but at once realize that neither $V_p(R)$ nor $\Omega_p(r)$ is affected by the gauge factor and thus may be computed as well on the basis of any local representation. $\Omega_p(r)$ is clearly the compensating term which accounts, as it were, for the stationary points of rank p situated at infinity, and it is easy to check this in the case of rational curves. Hence our formula is the *formula of the sought-for second main theorem*.

It is desirable, especially in view of later generalization to arbitrary Riemann surfaces, to prove it without recourse to the normalized representation. Proceeding along the same lines as in the rational case, one chooses a contragredient $A^p \neq 0$ for each $p = 0, 1, \ldots, n$ and applies the condenser formula to the quotient

$$Z^p(z) = \frac{(A^{p+1}X^{p+1})\cdot(A^{p-1}X^{p-1})}{(A^pX^p)^2}$$

with the result

$$V_p(R) + \{N_{p+1}(R;A^{p+1}) - 2N_p(R;A^p) + N_{p-1}(R;A^{p-1})\}$$
$$= [\tfrac{1}{2\pi} \int_0^{2\pi} \log |Z^p| \cdot d\vartheta]_{r_0}^R .$$

Adding the difference $[\ ]_{r_0}^R$ of

$$m_{p+1}(r;A^{p+1}) - 2m_p(r;A^p) + m_{p-1}(r;A^{p-1})$$

on either side and making use of the relation

$$N_p(R;A^p) + [m_p(r;A^p)]_{r_0}^R = T_p(R)$$

one again arrives at the formula (1.2).

The assurance that $\Omega_p(r)$ is the compensating term for the enumeration of stationary points of rank p, indeed the formula (1.2) itself, is of little value unless one can show $\Omega_p(r)$ to be negative, at least in some asymptotic sense for $r \to \infty$. Let us check whether that statement is correct for exponential curves of the simple type (II, 5.1). Such a curve has no stationary points of any rank; thus $V_p(R) = 0$. The expression for $T_p(R)$ is given by (II, 5.16), (II, 5.18), and hence what we have to examine is whether

$$L_{p+1} - 2L_p + L_{p-1} \leqq 0.$$

It this is so, the broken line which plots L_p as a function of $p(=0,1,\ldots,n)$ will be concave. We find for the first and second differences

$$L_p - L_{p-1} = \int_0^{2\pi} q_p(\vartheta)\, d\vartheta,$$

$$L_{p+1} - 2L_p + L_{p-1} = - \int_0^{2\pi} \{q_p(\vartheta)-q_{p+1}(\vartheta)\}\, d\vartheta,$$

and thus the second differences are indeed non-positive, because by definition

$$q_1(\vartheta) \geqq q_2(\vartheta) \geqq \ldots \geqq q_n(\vartheta).$$

The combined examples of the rational and the exponential curves leave little doubt that we have to expect an essentially negative $\Omega_p(r)$ even for arbitrary meromorphic curves. However, this will follow only after we have developed systematically the implications of a new idea which we now throw into the game, namely that of <u>averaging over all intersecting planes</u>.

§2. Average of the defect

The unitary sphere in complex n-space

$$(2.1) \qquad |\alpha_1|^2 + \ldots + |\alpha_n|^2 = 1,$$

is the ordinary unit sphere in the 2n-dimensional real space whose Cartesian coordinates are the real and imaginary parts of $\alpha_1, \ldots, \alpha_n$, and thus areas of parts of the sphere can be measured and continuous functions $f(\alpha)$ on the sphere be integrated. We use the notation $\int f(\alpha) \cdot d\omega_\alpha$, indicating by $d\omega_\alpha$ the infinitesimal area at the point α. The average $\mathfrak{M}_\alpha f(\alpha)$ of f is the quotient

$$\int f(\alpha) \cdot d\omega_\alpha / \int d\omega_\alpha.$$

Unitary transformations of the complex coordinates $\alpha_k = \alpha_k' + i\alpha_k''$ are orthogonal transformations of the 2n real coordinates α_k', α_k'', and hence the measure $d\omega_\alpha$ is invariant with respect to unitary transformations of the coordinates α_k.

We now propose to apply this process of averaging to the equation of the first main theorem for a given non-degenerate curve $\mathfrak{C}$:

$$(2.2) \qquad N(R;\alpha) + [m(r;\alpha)]_{r_0}^{R} = T(R).$$

In order to compute the integral over the α-sphere of

$$(2.3) \qquad m(r;\alpha) = \frac{1}{2\pi}\int_0^{2\pi} \log\frac{1}{\|\alpha x\|}\cdot d\vartheta$$

we prove the following basic

LEMMA 2. A. For every non-vanishing vector x the average of $\log\frac{1}{\|\alpha x\|}$ has the same value.

The proof is quite simple. We may assume $|x| = 1$ and then by a suitable unitary transformation change $x = (x_1,\ldots,x_n)$ into the unit vector $e = (1,0,\ldots,0)$. Because of unitary invariance the average

$$\mathfrak{M}_\alpha \log\frac{1}{\|\alpha x\|}$$

has the same value for x as for e, viz.

$$(2.4) \qquad \mathfrak{M}_\alpha \log\frac{1}{|\alpha_1|} = m_0.$$

By interchanging integration with respect to ϑ and (α) the definition (2.3) then gives

$$(2.5) \qquad \mathfrak{M}_\alpha m(r;\alpha) = m_0.$$

But that interchange has to be justified, and it was with this end in view that we developed in II, §4, the lemmas concerning the compensating function, in particular Lemma E with its corollary . As a continuous function $m(r;\alpha)$ is integrable over the α-sphere. Let $m_\delta(r;\alpha)$ have the same meaning as in II, §4. Then

$$\int m_\delta(r;\alpha)\, d\omega_\alpha = \frac{1}{2\pi}\int \left\{\int \log\frac{1}{\|\alpha x\|}\, d\vartheta\right\} d\omega_\alpha$$

with the inner integration extending over only that part

of the circle k_r where $\|\alpha x\| \geqq \delta$. Here, where we keep away from all singularities, the integrations are certainly interchangeable. The same argument by which Lemma 2. A has been proved shows that for all vectors $x \neq 0$ the integral

$$\int_{\|\alpha x\| \geqq \delta} \log \frac{1}{\|\alpha x\|} \cdot d\omega_\alpha = \int_{|\alpha_1| \geqq \delta} \log \frac{1}{|\alpha_1|} \cdot d\omega_\alpha$$

is independent of x. Let $m_0(\delta)$ denote the quotient of this number and the surface $\int d\omega_\alpha$ of the sphere. Then

$$\mathfrak{M}_\alpha m_\delta(r;\alpha) = m_0(\delta). \tag{2.6}$$

The improper integral defining m_0 is the limit of $m_0(\delta)$ for $\delta \to 0$, and since we know by Lemma II, 4. E that $m_\delta(r;\alpha)$ tends to $m(r;\alpha)$ uniformly on the entire α-sphere, (2.6) gives the desired result (2.5).

In view of (2.2) it implies the further equation

$$T(R) = \mathfrak{M}_\alpha N(R;\alpha). \tag{2.7}$$

We summarize:

THEOREM. The average of the defect $m(r;\alpha)$ for all possible positions of the intersecting plane (α) is independent of r and thus ~ 0. The order function $T(R)$ is the mean of $N(R;\alpha)$ and a function of R of regular type.

Indeed, being of regular type is a property which carries over from $N(R;\alpha)$ to its mean.

The theorem reveals that it is to be considered an extraordinary event when $N(R;\alpha)$ falls behind $T(R)$ by an appreciable percentage, and thus justifies the word <u>defect</u> for the difference $m(r;\alpha)$. The simplicity of our results is due to the fact that our metric is invariant under unitary transformations and that under this group all vectors x of unit length, $|x| = 1$, are equivalent.

In this respect the "circular" metric has a very decided advantage over all other metrics, in particular the "square" metric employed by R. Nevanlinna for the investigation of meromorphic functions. Expressions of T as averages of N for meromorphic functions were first developed by T. Shimizu (On the Theory of Meromorphic Functions, Jap. Jour. Math. 6, 1929), L. Ahlfors (Beiträge zur Theorie der meromorphen Funktionen, 7. Congr. Math. Scand., Oslo 1929, and [8]), and H. Cartan, C. R. Acad. Sci. Paris 189, 1929, 521-523 and 625-627.* The role of a metric in general and the unitary metric in particular for the study of meromorphic curves was fully realized in 1938 by H. and J. Weyl.

In order to familiarize ourselves with the process of integration over the unit sphere we now propose to compute the value of m_0. Suppose a continuous function $f(\alpha)$ to be given on the unit sphere and extended to the entire α-space by the condition of homogeneity

$$f(r\alpha_1,\ldots,r\alpha_n) = f(\alpha_1,\ldots,\alpha_n) \qquad \text{for } r > 0.$$

Then the average $\mathfrak{M}_\alpha f(\alpha)$ is equal to the quotient of the two integrals

$$(2.8)\quad \int f(\alpha)\cdot e^{-|\alpha_1|^2 - \ldots - |\alpha_n|^2}\cdot d\alpha_1' \, d\alpha_1'' \ldots d\alpha_n' \, d\alpha_n'',$$

$$(2.9)\quad \int e^{-|\alpha_1|^2 - \ldots - |\alpha_n|^2} \, d\alpha_1' \, d\alpha_1'' \ldots d\alpha_n' \, d\alpha_n''$$

extending over the whole space. Such integrals are easier to compute than integrals over the unit sphere. The factor $e^{-|\alpha|^2}$, rather than any other function of $|\alpha|^2$, has been chosen because of the simple functional equation

---

*Cf. also the general formulation by O. Frostman, 8. Congr. Math. Scand. Stockholm 1934, and Meddel. Lunds Univ. Mat. Sem. 3, 1935. A complete account in R. Nevanlinna, [10], Ch. VI, §§3-4.

$$e^{-(t_1+t_2)} = e^{-t_1} \cdot e^{-t_2}.$$

(2.9) is readily evaluated. When expressed in terms of polar coordinates for each complex variable α_k,

$$\alpha_k = \rho_k \cdot e^{i\phi_k} \qquad (\rho_k \geqq 0,\ 0 \leqq \phi_k < 2\pi),$$

$$d\alpha_k' \, d\alpha_k'' = \rho_k d\rho_k \cdot d\phi_k = \tfrac{1}{2} dt_k \cdot d\phi_k \text{ with } t_k = \rho_k^2,$$

it becomes

$$\pi^n \int_0^\infty \cdots \int_0^\infty e^{-t_1-\cdots-t_n} \cdot dt_1 \ldots dt_n,$$

and since $\int_0^\infty e^{-t} dt = 1$, the value π^n results for (2.9).

$2m_0$ is the average of $- \log (|\alpha_1|^2/|\alpha|^2)$. Consider any positive continuous function $g(t)$ defined for $0 < t \leqq 1$. We maintain:

LEMMA 2. B.

(2.10) $$\frac{1}{\pi^n} \int g\left(\frac{|\alpha_1|^2}{|\alpha|^2}\right) \cdot e^{-|\alpha|^2} d\alpha_1' \, d\alpha_1'' \ldots d\alpha_n' \, d\alpha_n'' = (n-1) \int_0^1 g(t)(1-t)^{n-2} dt,$$

the convergence of either of the two integrals implying that of the other.

Proof. By the introduction of polar coordinates for each α_k the left member of (2.10) changes into

(2.11) $$\int_0^\infty \cdots \int_0^\infty g\left(\frac{t_1}{t_1+\cdots+t_n}\right) \cdot e^{-(t_1+\cdots+t_n)} dt_1 \ldots dt_n.$$

The reversible substitution

$$t_1 = \tau(s_2+\cdots+s_n),\ t_2 = (1-\tau)s_2,\ \ldots,\ t_n = (1-\tau)s_n$$

$$(t_1+\cdots+t_n = s_2+\cdots+s_n)$$

with the Jacobian

$$(s_2+\dots+s_n)(1-\tau)^{n-2}$$

carries (2.11) into

$$\int_0^1 g(\tau)(1-\tau)^{n-2}d\tau\cdot\int_0^\infty\dots\int_0^\infty (s_2+\dots+s_n)e^{-(s_2+\dots+s_n)}ds_2\dots ds_n$$

Because of

$$\int_0^\infty e^{-s}ds = 1 \quad\text{and}\quad \int_0^\infty s\cdot e^{-s}\, ds = 1$$

each individual term of the (n-1)-fold integral, like

$$\int_0^\infty\dots\int_0^\infty s_2\cdot e^{-(s_2+\dots+s_n)}ds_2\dots ds_n,$$

equals 1, and thus the value

$$(n-1)\int_0^1 g(\tau)(1-\tau)^{n-2}d\tau$$

is obtained for (2.11) as stated.

For $g(t) = -\log t$ we transform the result

$$(n-1)\int_0^1 g(t)(1-t)^{n-2}dt = -(n-1)\int_0^1 \log(1-t)\cdot t^{n-2}dt$$

$$= -\int_0^1 \log(1-t)\cdot d(t^{n-1}-1)$$

through integration by parts into

$$\int_0^1 \frac{t^{n-1}-1}{t-1}\,dt = \int_0^1 (1+\dots+t^{n-2})\,dt = \frac{1}{1}+\frac{1}{2}+\dots+\frac{1}{n-1},$$

and thus arrive at the desired evaluation,

$$m_0 = \frac{1}{2}\left(\frac{1}{1}+\frac{1}{2}+\dots+\frac{1}{n-1}\right). \tag{2.12}$$

§3. On integration in general

When integrating over a topological space, especially over a compact space, one frequently finds oneself forced to subdivide that space into a number of pieces, each referred to its own system of coordinates. However, the boundary lines may cause trouble, which can be avoided if, instead of cutting up the manifold, one covers it with "shingles", elementary pieces which are allowed to overlap. There is a general tendency in topology today to replace dissection by coverage.

Let us suppose that in some way a class of "admissible coordinate systems" for our (oriented) space is circumscribed, with the following properties. For any point $\mathfrak{p}_0$ there is an admissible system of (real) coordinates $x_1, \ldots, x_n$ mapping an open chunk $\mathfrak{N}(\mathfrak{p}_0)$ of the space topologically on a parallelotope

$$-l_1' < x_1 < l_1', \ldots, -l_n' < x_n < l_n' \qquad (l_i' > 0)$$

such that $\mathfrak{p}_0$ is mapped into the origin. For any point $\mathfrak{p}_1$ in $\mathfrak{N}(\mathfrak{p}_0)$ the quantities $x_i - x_i^1$ ($x_i^1 = x_i(\mathfrak{p}_1)$) constitute an admissible coordinate system. A part P of $\mathfrak{N}(\mathfrak{p}_0)$, which is mapped by our coordinates upon a closed parallelotope

$$(3.1) \qquad -l_i \leqq x_i \leqq l_i \quad \text{with} \quad 0 < l_i < l_i' \quad (i=1,\ldots,n)$$

is said to be a <u>block</u> of center $\mathfrak{p}_0$ (and coordinates x_i). If $x_1^*, \ldots, x_n^*$ are any admissible coordinates for the same point $\mathfrak{p}_0$ the transformation functions $x_i^*(x_1,\ldots,x_n)$ connecting the two systems in the neighborhood of $\mathfrak{p}_0$ have continuous first derivatives $(\partial x_i^*)/(\partial x_j)$ with a <u>positive</u> Jacobian in the neighborhood of the origin $(x_1,\ldots,x_n) = (0,\ldots,0)$.

A <u>function</u> ψ defined in the neighborhood of $\mathfrak{p}_0$ is expressible in terms of the coordinates x_i, $\psi = f(x_1,\ldots,x_n)$ and its expressions f and f^* in terms of two admissible coordinate systems x_i, x_i^* are related by the equation $f^*(x_1^*,\ldots,x_n^*) = f(x_1,\ldots,x_n)$ in which x_i^* is to be replaced by $x_i^*(x_1,\ldots,x_n)$. A <u>density</u> Ψ at the point $\mathfrak{p}_0$ is given if there is a number F associated with any admissible system of coordinates for $\mathfrak{p}_0$ such that the values F, F^* for two such systems x_i, x_i^* are related by

$$(3.2_0) \qquad F^* \cdot \det\left(\frac{\partial x_i^*}{\partial x_j}\right)_0 = F.$$

The notion is very similar to that of a differential. Let a density Ψ be defined for all points $\mathfrak{p}_1$ in a certain neighborhood of $\mathfrak{p}_0$ which is covered by the admissible coordinate system x_i, and set $x_i(\mathfrak{p}_1) = x_i^1$. Then Ψ will have a definite value $F(x_1^1,\ldots,x_n^1)$ at $\mathfrak{p}_1$ with respect to the coordinates $x_i - x_i^1$, a fact which we indicate by the equation

$$\Psi(\mathfrak{p}_1) = F(x_1^1,\ldots,x_n^1) \text{ or } \quad \Psi(\mathfrak{p}) = F(x_1,\ldots,x_n).$$

The relation (3.2_0) carries over to all points $\mathfrak{p}$ sufficiently near $\mathfrak{p}_0$:

$$(3.2) \quad F^*(x_1^*,\ldots,x_n^*) \cdot \det\left(\frac{\partial x_i^*}{\partial x_j}\right) = F(x_1,\ldots,x_n).$$

Ψ is said to be continuous at $\mathfrak{p}_0$ if its value $F(x_1,\ldots,x_n)$ is continuous at the origin. Because of (3.2) this condition is independent of the coordinate system. The product of a density and a function is a density.

To every continuous density Ψ defined on the whole manifold and vanishing outside a compact part of it, we can assign an integral $\int\Psi$ which satisfies the following two axioms:

(i) $\int(\Psi_1 + \Psi_2) = \int\Psi_1 + \int\Psi_2$.

(ii) If Ψ vanishes outside a block P, (3.1), while having the value $F(x_1,\ldots,x_n)$ inside the block with respect to its coordinates x_i, then $\int\Psi$ is identical with the Riemann integral

$$\int_{-l_n}^{l_n} \cdots \int_{-l_1}^{l_1} F(x_1,\ldots,x_n)\, dx_1 \ldots dx_n.$$

Proof. Consider only those continuous Ψ which vanish outside a given compact part G of the manifold. Enclose each point $\mathfrak{p}_0$ in a block $P(\mathfrak{p}_0) = P$ represented in terms of its coordinates x_i by (3.1). Let θ be a fixed positive number < 1. The block P contains the smaller block P^θ defined by $-\theta l_i \leqq x_i \leqq \theta l_i$. Because G is compact we can ascertain a finite number of points $\mathfrak{p}_q$ $(q = 1,2,\ldots)$ in G such that the contracted blocks $P^\theta(\mathfrak{p}_q)$ cover G. Denote by $x_i^{(q)}$ the coordinates corresponding to $P_q = P(\mathfrak{p}_q)$. As we shall presently see, we may further construct continuous functions μ_q ("Dieudonné factors") enjoying the following properties: $0 \leqq \mu_q \leqq 1$ everywhere, $\mu_q = 0$ outside P_q, and

$$\sum \mu_q = 1 \quad \text{in G.}$$

Given a continuous density Ψ which vanishes outside G we may break Ψ into the sum

$$(3.3) \qquad \Psi = \sum \Psi_q \qquad \Psi_q = \mu_q \Psi$$

(Dieudonné partition). Notice that (3.3) holds also for points outside G because Ψ vanishes there, and that the individual summand Ψ_q vanishes outside P_q. Compute the integral of Ψ_q according to axiom (ii) by means of the coordinates $x_i^{(q)}$, and in compliance with (i) form

$$\int \Psi = \sum \int \Psi_q \, .$$

Vice versa, the functional $\int \Psi$ thus computed for any continuous density Ψ vanishing outside G obeys the two laws (i) and (ii). The first is evident, the second follows from the law of transformation (3.2), which implies that for a Ψ vanishing outside P, the integral of $\mu_q \Psi$ turns out to be the same whether computed by means of the coordinates $x_i^{(q)}$ belonging to P_q, or the coordinates x_i belonging to P. As we have seen, the rules (i) and (ii) uniquely determine the value $\int \Psi$, and hence that value is of necessity independent of the auxiliary blocks, coordinates, and Dieudonné partition by which it has been computed above.

It remains to construct the functions μ_q. A function $\lambda(\mathfrak{p})$ which satisfies the inequality $0 \leqq \lambda(\mathfrak{p}) \leqq 1$ everywhere is said to be a probability function. Given an admissible block P, (3.1), we can easily construct a continuous probability function $\lambda(\mathfrak{p})$ which equals 1 in the shrunk P^θ and vanishes outside P. Take a continuous probability function $\lambda(x)$ of the real variable x which has the constant value 1 for $-\theta \leqq x \leqq \theta$ while it vanishes outside the interval $(-1,1)$ and set

$$\lambda(\mathfrak{p}) = \lambda\left(\frac{x_1}{l_1}\right) \cdots \lambda\left(\frac{x_n}{l_n}\right) \qquad \text{inside P.}$$

If we wish we can make $\lambda(\mathfrak{p})$ as smooth as we like, in particular take care that it has continuous first derivatives (is of class D_1). We simply have to let the basic function $\lambda(x)$ be of class D_1. In this way we construct a function $\lambda_q(\mathfrak{p})$ for each of our covering blocks P_q and then argue as follows.

a, b being the probabilities of two statistically independent events ($0 \leqq a, b \leqq 1$) the one or the other event will happen with the probability

$$a \vee b = a + b - ab.$$

Notice the inequalities

$$a \vee b \leqq 1, \quad a \vee b \geqq a, \geqq b, \text{ \underline{a fortiori} } \geqq 0,$$

and the fact that the probability sum $a \vee b$ may be written as an ordinary sum $a + b'$ with $b' = b - ab$, $0 \leqq b' \leqq b$. The probability sum $\lambda_1 \vee \lambda_2 \vee \ldots$ of our probability functions λ_q associated with the blocks P_q equals 1 everywhere in G because at any point of G at least one of the $\lambda_q = 1$. Write it in the form of an ordinary sum $\mu_1 + \mu_2$

+ ... with

$$\mu_1 = \lambda_1, \qquad \mu_q = \lambda_q - \lambda_q(\lambda_1 \vee \dots \vee \lambda_{q-1}),$$

and thus complete the construction of the Dieudonné factors μ_q.*

Computation of an integral by dissecting the domain of integration into disjoint pieces G_q, each covered by a coordinate system of its own, would mean decomposition of the density Ψ to be integrated, $\Psi = \sum \sigma_q \Psi$, by means of the characteristic functions σ_q of the parts G_q:

$$\sigma_q = 1 \text{ inside}, \qquad = 0 \text{ outside } G_q.$$

In the Dieudonné procedure these discontinuous σ_q are replaced by the smooth μ_q. Its advantage for the integration of continuous functions is obvious.

In integrals extending over parts of a plane with the real coordinates x, y it is customary to write the element of integration as dx dy; this corresponds to the idea that the plane is cut up into infinitesimal rectangles of sides dx and dy parallel to the axes of coordinates. However, when dealing with a complex variable $z = x + iy$ and analytic transformations, it is more reasonable to write $dz\,d\bar{z} = |dz|^2$ instead; this is the area of an infinitesimal square of arbitrary orientation spanned by the vectors dz and $i \cdot dz$. Indeed, conformal transformation carries an infinitesimal square into a square, but in general the orientation of rectangles parallel to the real and imaginary axes is destroyed. Our form of writing is suggestive of the law of transformation of integrals

$$(3.4) \qquad \int \phi \, dw \, d\bar{w} = \int \phi \, \left|\frac{dw}{dz}\right|^2 dz \, d\bar{z}$$

under a one-to-one conformal transformation $w = w(z)$.

*Sometimes it is convenient to use spheres instead of parallelotopes, but the definition of a multiple Riemann integral is simpler for parallelotopes.

If $w(z)$ is any regular analytic function in a compact part G of the z-plane of definite Jordan area, the mapping $z \to w$ effected by it is generally speaking not one-to-one. But it may be made into a one-to-one mapping if the w-plane is replaced by a suitable covering Riemann surface (G is turned into it by the convention that z_0 covers the point w_0 of the w-plane if $w(z_0) = w_0$). The area of this Riemann surface will then be given by the integral

$$\int_G \left|\frac{dw}{dz}\right|^2 dz\, d\bar{z}.$$

If the Riemann surface has $n(w)$ points over the point w of the w-plane we thus obtain the equation

$$(3.5) \qquad \int n(w)\,dw\, d\bar{w} = \int_G \left|\frac{dw}{dz}\right|^2 dz\, d\bar{z}.$$

In determining areas and integrating over limited parts of the manifold we have been unfaithful to our principle of integrating only continuous functions over the whole manifold, to which Dieudonné's procedure is adapted. We return ruefully to it by introducing a continuous function $\phi(z)$ in the whole z-plane vanishing outside a compact part G. The function $w(z)$ is supposed to be regular in G, but G need not have a definite Jordan area. Denoting by $\nu(z_0;w_0)$ the order in which $w(z) - w_0$ vanishes for $z = z_0$ and setting

$$N_\phi(w) = \sum_z \phi(z)\cdot \nu(z;w)$$

we find instead of (3.5) the formula

$$(3.6) \qquad \int N_\phi(w)\,dw\, d\bar{w} = \int \phi(z) \left|\frac{dw}{dz}\right|^2 dz\, d\bar{z}.$$

This improvement of (3.4) can be neatly demonstrated by the Dieudonné partition as follows. (Continuity of $N_\phi(w)$ will result as a byproduct of the proof.)

Let z_0 be any point of G, $w(z_0) = w_0$, and suppose

$w(z) - w_0$ to vanish at $z = z_0$ to the order $\nu(z_0; w_0) = \nu > 0$. Locally the solution of the equation $w(z) = w$ may be obtained in the form

$$w - w_0 = t^{\nu}, \qquad t = b_1(z-z_0) + \ldots \qquad (b_1 \neq 0).$$

In a certain circle $K(z_0)$, $|z-z_0| < \rho_0$, the function $t(z)$ will have an analytic inverse and $z \to t$ will map $K(z_0)$ one-to-one conformally upon a certain region in the t-plane. Choose a positive $\theta < 1$ and form the contracted circle $K^{\theta}(z_0)$, $|z-z_0| < \theta\rho_0$. Since G may be covered by a finite number of such contracted circles $K^{\theta}(z_0)$, the Dieudonné partition reduces the proof to the case where the continuous $\phi(z)$ vanishes outside $K(z_0)$. Then by means of the one-to-one mapping $z \rightleftarrows t$ and with $\phi(z(t)) = \phi^*(t)$,

$$\int \phi(z) \left|\frac{dw}{dz}\right|^2 dz\, d\bar{z} = \int \phi^*(t) \left|\frac{dw}{dt}\right|^2 dt\, d\bar{t}.$$

Introduce polar coordinates, $t = re^{i\vartheta}$, and the latter integral will easily be changed into the integral $\int N_{\phi}(w) dw\, d\bar{w}$ where $N_{\phi}(w)$ is the sum of the values of ϕ^* at the ν roots $t = t_1, \ldots, t_{\nu}$ of the equation $t^{\nu} = w - w_0$.

The formula

$$(3.7) \quad \int N_{\phi}(w)\psi(w) dw\, d\bar{w} = \int \phi(z)\psi(w(z)) \left|\frac{dw}{dz}\right|^2 dz\, d\bar{z}$$

involving an arbitrary continuous function $\psi(w)$ of w is a trivial generalization of (3.6). One has simply to replace $\phi(z)$ by $\phi(z)\cdot\psi(w(z))$.

This is the service which Dieudonné's partition always performs: reducing a global to the corresponding local statement about an integral.

When dealing with integration over several complex variables $z_1, \ldots, z_n$ we write the element of integration as $dz_1\, d\bar{z}_1 \ldots dz_n\, d\bar{z}_n$. What is its law of transformation under an analytic transformation $w_i = w_i(z_1, \ldots, z_n)$ of the variables? Their differentials undergo a linear transformation with the coefficients $(\partial w_i)/(\partial z_k)$. Consider therefore any linear transformation

$$w_i = \sum_k a_{ik} z_k$$

and denote its determinant det (a_{ik}) by Δ. Splitting into real and imaginary parts, $z_k = x_k + iy_k$, $w_k = u_k + iv_k$, we want to know what the determinant of the linear transformation is which carries the 2n variables x_k, y_k into the 2n variables u_k, v_k. Instead of $x = \frac{1}{2}(z+\bar{z})$, $y = \frac{1}{2i}(z-\bar{z})$ we may use z, $\bar{z}$ and then see at once that that determinant equals $\Delta\bar{\Delta}$. This result finds its expression in the general formula for analytic transformations

$$dw_1\, d\bar{w}_1 \ldots dw_n\, d\bar{w}_n = \Delta\bar{\Delta} \cdot dz_1\, d\bar{z}_1 \ldots dz_n\, d\bar{z}_n,$$

$$\Delta = \det\left(\frac{\partial w_i}{\partial z_k}\right).$$

Observe that the factor $\Delta\bar{\Delta} = |\Delta|^2$ is never negative.

§4. Average of the N-term

We are now well prepared to integrate the term $N(R;\alpha)$ over the unit sphere in α-space. Indeed, the formula (3.6) refers exactly to such a quantity as $N(R;\alpha)$.

Let $x_i = x_i(z)$ be a reduced representation of our non-degenerate curve $\mathfrak{C}$ for the neighborhood of the point $z = z_0$, and $x_1(z_0) \neq 0$. The points z of intersection with the plane (α) are determined by the equation

$$(4.1) \qquad \alpha_1 \cdot x_1(z) + \ldots + \alpha_n \cdot x_n(z) = 0$$

which for points z sufficiently near to z_0 may be written in the form

$$(4.2) \qquad \alpha_1 = -\frac{\alpha_2\, x_2(z) + \ldots + \alpha_n\, x_n(z)}{x_1(z)}.$$

For fixed $\alpha_2, \ldots, \alpha_n$ the right-hand side is a function $\alpha_1(z)$. Its derivative is computed from (4.1):

$$x_1 \frac{d\alpha_1}{dz} + (\alpha_1 x_1' + \ldots + \alpha_n x_n') = 0, \quad \frac{d\alpha_1}{dz} = -\frac{(x'\alpha)}{x_1}.$$

Apply the local formula (3.7) by letting α_1 play the role of w and choosing $\psi(\alpha_1) = \exp(-|\alpha_1|^2)$:

$$(4.3) \quad \int N_\phi(\alpha)\, e^{-|\alpha_1|^2} d\alpha_1\, d\bar{\alpha}_1 = \int \phi(z) \frac{|(x'\alpha)|^2}{|x_1|^2} e^{-|\alpha_1|^2} dz\, d\bar{z}.$$

Here $\phi(z)$ is any continuous function vanishing outside a certain circle $|z-z_0| < \rho_0$ around z_0, and

$$N_\phi(\alpha) = \sum\nolimits_z \phi(z)\nu(z;\alpha)$$

with $\nu(z_0;\alpha)$ as in II, §2 denoting the order of vanishing at $z = z_0$ of the linear form (4.1). Consequently

$$(4.4) \quad \begin{aligned} &\int \cdots \int N_\phi(\alpha) \cdot e^{-|\alpha_1|^2 - \ldots - |\alpha_n|^2} d\alpha_1\, d\bar{\alpha}_1 \ldots d\alpha_n\, d\bar{\alpha}_n = \\ &\int \phi(z) \left\{ \int \cdots \int e^{-|\alpha_1|^2 - \ldots - |\alpha_n|^2} \frac{|(x'\alpha)|^2}{|x_1|^2} d\alpha_2\, d\bar{\alpha}_2 \ldots d\alpha_n\, d\bar{\alpha}_n \right\} dz\, d\bar{z} \end{aligned}$$

The left side equals $\pi^n \mathfrak{M}_\alpha N_\phi(\alpha)$. In the inner integral on the right side α_1 stands for the quantity computed from (4.2) in terms of $z;\ \alpha_2, \ldots, \alpha_n$. Let $x = (x_1, \ldots, x_n)$ be a given vector, $x_1 \neq 0$, and $d\sigma$ an infinitesimal area of the plane defined in α-space by the equation $x_1\alpha_1 + \ldots + x_n\alpha_n = 0$ and $d\alpha_2\, d\bar{\alpha}_2 \ldots d\alpha_n\, d\bar{\alpha}_n$ its perpendicular projection on the coordinate plane $\alpha_1 = 0$. Then we have the elementary geometric formula

$$(4.5) \qquad d\sigma = \frac{|x|^2}{|x_1|^2} d\alpha_2\, d\bar{\alpha}_2 \ldots d\alpha_n\, d\bar{\alpha}_n.$$

Therefore we may write the inner integral at the right side of (4.4) as

$$(4.6) \qquad \int\cdots\int_{(x\alpha)=0} e^{-|\alpha_1|^2 - \cdots - |\alpha_n|^2} \cdot \frac{|(x'\alpha)|^2}{|x|^2}\, d\sigma.$$

The formula then holds independently of the restriction $x_1(z_0) \neq 0$. It will presently be shown that for any two vectors $x \neq 0$ and x' the integral (4.6) has the value

$$\pi^{n-1} \frac{|[xx']|^2}{|x|^4}.$$

If we now again interpret x_i as $x_i(z)$ and x_i' as its derivative, the expression

$$(4.7) \qquad S(z) = 2\, \frac{|[xx']|^2}{|x|^4}$$

is clearly independent of the gauge factor, and according to Dieudonné's procedure the resulting formula

$$(4.8) \qquad \mathfrak{M}_\alpha N_\phi(\alpha) = \frac{1}{2\pi}\int \phi(z) S(z)\, dz\, d\bar{z}$$

holds not only locally but also globally, i. e. for any continuous function $\phi(z)$ vanishing outside a bounded part of the z-plane.

Before proceeding further we ought to complete the argument by proving (4.5) and evaluating (4.6). In (4.5) assume $|x| = 1$. Ascertain a unitary transformation $\|x_{ij}\|$ of which $x_{11} = x_1 \neq 0, \ldots, x_{1n} = x_n$ is the first row:

$$\sum_j x_{ij}\alpha_j = \beta_i \qquad (i,j = 1,\ldots,n).$$

If α_1 is determined as function of $\alpha_2, \ldots, \alpha_n$ by $\beta_1 = 0$, then the determinant D of the substitution $(\alpha_2,\ldots,\alpha_n) \rightarrow (\beta_2,\ldots,\beta_n)$ is

$$D = \left| x_{ij} - \frac{x_{i1}x_{1j}}{x_1} \right| \quad i,j = 2, \ldots, n.$$

$x_1 D$ arises from

$$\begin{vmatrix} x_{11}, & x_{12}, & \dots, & x_{1n} \\ x_{21}, & x_{22}, & \dots, & x_{2n} \\ \cdot & \cdot & \cdot & \cdot \\ x_{n1}, & x_{n2}, & \dots, & x_{nn} \end{vmatrix}$$

by subtracting from the 2nd, ..., n^{th} columns the first column multiplied by x_{12}/x_1, ... , x_{1n}/x_1 respectively. Hence

$$D = \frac{1}{x_1} \cdot \det (x_{ij})_{i,j = 1,\dots,n}$$

and $D\overline{D} = \frac{1}{x_1\overline{x}_1}$, or

$$d\beta_2\, d\overline{\beta}_2 \dots d\beta_n\, d\overline{\beta}_n = \frac{1}{|x_1|^2}\, d\alpha_2 d\overline{\alpha}_2 \dots\ d\alpha_n d\overline{\alpha}_n,$$

as we had maintained.

In verifying the equation

$$\underset{(x\alpha)=0}{\int\dots\int} e^{-|\alpha_1|^2 - \dots - |\alpha_n|^2} \cdot \frac{|(x'\alpha)|^2}{|x|^2}\, d\sigma = \pi^{n-1}\, \frac{|[xx']|^2}{|x|^4}$$

involving two vectors $x \neq 0$ and x' we may use a unitary coordinate system adapted to these vectors in such a manner that they assume the form

$$\begin{aligned} x\ &= (x_1,\ 0,\ 0,\ \dots,\ 0), \\ x' &= (x_1',\ x_2',\ 0,\ \dots,\ 0). \end{aligned}$$

Then

$$|x| = |x_1|, \quad |[xx']| = |x_1 x_2'|,$$

and the integral (4.6) becomes

$$\frac{|x_2'|^2}{|x_1|^2} \int\dots\int |\alpha_2|^2 e^{-|\alpha_2|^2 - \dots - |\alpha_n|^2} \cdot d\alpha_2\, d\overline{\alpha}_2 \dots d\alpha_n\, d\overline{\alpha}_n.$$

The numerical coefficient

$$\pi^{n-1}\int_0^{\infty}\cdots\int_0^{\infty} t_2 e^{-(t_2+\cdots+t_n)}\,dt_2 \ldots dt_n = \pi^{n-1}$$

has been computed before; see the proof of Lemma 2. B.

The expression S, (4.7), has a simple geometric significance. The square of the distance of any two infinitely near points x and x + dx on the curve, dx = x'dz, is given by

$$(4.9)\quad \|x:x+dx\|^2 = \frac{|[x,dx]|^2}{|x|^4} = \frac{|[xx']|^2}{|x|^4}dz\,d\bar{z} = \tfrac{1}{2}S(z)\cdot dz\,d\bar{z}.$$

This form of the "line element" for the manifold $\mathfrak{C}$ of two (real) dimensions reveals that the z-plane is mapped conformally onto the curve, and hence the integral $\frac{1}{2}\int S(z)dz\,d\bar{z}$ extending over any part of the z-plane represents the area of the corresponding part of $\mathfrak{C}$ while

$$\tfrac{1}{2}\int\phi(R;z)S(z)\,dz\,d\bar{z}$$

could be described as the valence of the circular area $|z| \leqq R$ on the curve. Here, as in Ch. II, $\phi(R;z)$ denotes the potential of the circular condenser of radii r_0 and R.

We now substitute this potential for ϕ in (4.8). Combining the result with (2.7) we arrive at a proposition which reveals an entirely new aspect of the order function:*

THEOREM. $\pi T(R)$ is the valence of the circular area $|z| \leqq R$ of the curve $\mathfrak{C}$, in formula

$$(4.10)\quad T(r) = \frac{1}{2\pi}\int\phi(R;z)S(z)dz\,d\bar{z} \text{ with } S(z) = 2\frac{|[xx']|^2}{|x|^4}.$$

---

*The paper [11] by H. and J. Weyl contains the formula in logarithmized form only, as it is needed to establish the connection between T and Ω (see §6); the exact theorem as given here and its generalization to arbitrary rank p (§5) are due to L. Ahlfors [12]. For another approach see J. Dufresnoy, C. R. Acad. Sci. Paris 211, 1940, 536-538 and 628-631.

Incidentally this theorem once more proves that T is a function of regular type.

Our chief interest is in non-degenerate curves. Even so, removal of the hypothesis of non-degeneracy has its value because the associated curve $\mathfrak{C}_p$ in the $\binom{n}{p}$-space $\mathfrak{S}^{(p)}$ of all p-ads might well be degenerate. Observe then that for degenerate (though not totally degenerate) curves (II, 4.8) proves the average of $m(r;\alpha)$ to be independent of r because that of $'m(r;\alpha)$ is. Thus (2.7) carries over. The computation of the average of $N(R;\alpha)$ in this section has nowhere resorted to the hypothesis of non-degeneracy.

§5. The order of rank p

THEOREM. The order $T_p(R)$ of rank p (=1,...,n-1) of a non-degenerate meromorphic curve $\mathfrak{C}$ may be expressed as a linear combination of $\phi(R;z)$ over z with positive weights $S^p(z)$,

$$(5.1) \qquad T_p(R) = \frac{1}{2\pi}\int \phi(R;z) S^p(z)\,dz\,d\bar{z},$$

the weight being given by

$$(5.2) \qquad S^p(z) = 2\,\frac{|X^{p-1}|^2 \cdot |X^{p+1}|^2}{|X^p|^4}.$$

Proof. If $\mathfrak{C}_p$ is not totally degenerate, then (4.10) gives

$$T_p(R) = \frac{1}{\pi}\int \phi(R;z)\,\frac{|\{X^p, \dot{X}^p\}|^2}{|X^p|^4}\,dz\,d\bar{z}.$$

We operate in the $\binom{n}{p}$-space $\mathfrak{S}^{(p)}$ in which X^p, $\dot{X}^p = \frac{dX^p}{dz}$ are two vectors and $\{X^p, \dot{X}^p\}$ designates their skew-symmetric product with the components

$$X^p(i_1 \dots i_p)\dot{X}^p(j_1 \dots j_p) - X^p(j_1 \dots j_p)\dot{X}^p(i_1 \dots i_p)$$

$$(i_1 < \dots < i_p, \quad j_1 < \dots < j_p).$$

(We cannot denote that product by the symbol $[X^p, \dot{X}^p]$ which has already been given a different meaning.) Observe that

$$(5.3)\quad X^p = [x, x', \ldots, x^{(p-1)}],\ \dot{X}^p = [x, x', \ldots, x^{(p-2)}, x^{(p)}].$$

In order to compute

$$|\{X^p, \dot{X}^p\}|$$

for any given vectors $x, x', \ldots, x^{(p)}$ we adapt our unitary coordinate system to these vectors in such a way that they assume the form

$$(5.4)\quad \begin{aligned} x &= (x_1, \quad 0, \quad 0, \ldots, 0), \\ x' &= (x_1', \quad x_2', \quad 0, \ldots, 0), \\ &\vdots \\ x^{(p)} &= (x_1^{(p)}, x_2^{(p)}, \ldots, x_{p+1}^{(p)}, 0, \ldots, 0). \end{aligned}$$

Then the only non-vanishing ordered component of X^p is

$$X^p(1,2,\ldots,p) = x_1 x_2' \ldots x_p^{(p-1)}$$

whereas the only two non-vanishing ordered components of $\dot{X}^p$ are

$$\dot{X}^p(1,\ldots,p) = x_1 x_2' \ldots x_{p-1}^{(p-2)} x_p^{(p)},$$

$$\dot{X}^p(1,\ldots,p-1,p+1) = -x_1 x_2' \ldots x_{p-1}^{(p-2)} x_{p+1}^{(p)} .$$

Remembering that for any two vectors of the form

$$X = (X_1, 0, 0, \ldots, 0),\ \dot{X} = (\dot{X}_1, \dot{X}_2, 0, \ldots, 0)$$

the absolute value $|\{X, \dot{X}\}|$ equals $|X_1 \dot{X}_2|$, we find

$$\begin{aligned} |\{X^p, \dot{X}^p\}| &= |x_1 x_2' \ldots x_p^{(p-1)} \cdot x_1 x_2' \ldots x_{p-1}^{(p-2)} x_{p+1}^{(p)}| \\ &= |x_1 x_2' \ldots x_{p-1}^{(p-2)}| \cdot |x_1 x_2' \ldots x_p^{(p-1)} x_{p+1}^{(p)}| \end{aligned}$$

which in invariant form $= |X^{p-1}| \cdot |X^{p+1}|$. Thus the fundamental equations (5.1), (5.2) have been shown to hold whenever at least p of the coordinates $x_1, \ldots, x_n$ of the generic point on the curve $\mathfrak{C}$ are linearly independent, in particular if $\mathfrak{C}$ is non-degenerate.

In the latter case the curve projected from $\{A^h\}$ is also non-degenerate whatever the h-element $\{A^h\}$. Thus

$$T_p(R;A^h) = \frac{1}{2\pi}\int \phi(R;z) S^p(z;A^h) dz\, d\bar{z},$$

$$(5.5) \qquad S^p(z;A^h) = 2\,\frac{|[A^h X^{p-1}]|^2 \cdot |[A^h X^{p+1}]|^2}{|[A^h X^p]|^4},$$

$$(h \geqq 0,\ p \geqq 1,\ p+h \leqq n-1).$$

In this proof we have done what up to now we had carefully avoided: we have operated in the $\binom{n}{p}$-space of the general p-ads. The formula (5.1) has actually been proved by integrating over the manifold of all p-ads A^p of modulus 1. Returning to the hypothesis of a non-degenerate $\mathfrak{C}$ we shall now give a second demonstration which does not break the rules of conduct that have guided us so far.

First we apply (4.10) to the dual curve with the coordinates $\xi_1, \ldots, \xi_n$:

$$T_{n-1}(R) = \frac{1}{\pi}\int \phi(R;z)\,\frac{|[\xi\xi']|^2}{|\xi|^4}\, dz\, d\bar{z}.$$

But

$$*W = X^n, \quad *\xi = X^{n-1}, \quad *[\xi\xi'] = W \cdot X^{n-2}$$

where W stands for the Wronskian of $\mathfrak{C}$. Consequently

$$\frac{|[\xi\xi']|^2}{|\xi|^4} = \frac{|X^n|^2 \cdot |X^{n-2}|^2}{|X^{n-1}|^4}.$$

Thus the equation (5.1) for $p = n-1$,

$$T_{n-1}(R) = \frac{1}{2\pi}\int \phi(R;z) S^{n-1}(z)\, dz\, d\bar{z}$$

results.

Next we apply this equation to the projection of $\mathfrak{C}$ from $\{A^h\}$, which is a curve in an (n-h)-space. Setting $n - h - 1 = p$ we get

$$T_p(R;A^h) = \frac{1}{2\pi}\int \phi(R;z)S^p(z;A^h)dz\, d\bar{z},$$

which is formula (5.5) for $h = n - p - 1$. From here we work down by induction and, assuming p to be a given rank number ($1 \leqq p \leqq n - 1$), thus obtain one after the other the formulas (5.5) for $h = n - p - 1, n - p - 2, \ldots, 0$. The last equation for $h = 0$ coincides with the desired result (5.1).

In carrying out this program we use the fundamental relation for projection, Ch. II, (7.3),

$$(5.6)\quad T_p(R) - T_p(R;A^h) = \tilde{N}_p(R;A^h) + [\tilde{m}_p(r;A^h)]_{r_0}^{R}.$$

Let A^{h-1} be a given special (h-1)-ad normalized by $|A^{h-1}| = 1$, and generate $\{A^h\}$ from $\{A^{h-1}\}$ by adjunction of a vector a of length 1 perpendicular to $\{A^{h-1}\}$, $A^h = [a,A^{h-1}]$. Then also $|A^h| = 1$. We keep A^{h-1} fixed, but let a vary over all the vectors of the nature just described, and average with respect to a over the unit sphere in the (n-h+1)-space $\{\sim A^{h-1}\}$, a process which we indicate by $\underset{A^h \supset A^{h-1}}{\mathfrak{M}}$. Identifying $\{A^{h-1}\}$ with the subspace spanned by the first h - 1 members of a normal vector basis, we interpret $[A^{h-1}X^p]$ as the perpendicular projection $\tilde{X}^p$ of X^p upon the space $\{\sim A^{h-1}\}$ and then write $[A^hX^p] = [a\tilde{X}^p]$. The projection $\tilde{X}^p = [\tilde{x},\tilde{x}',\ldots,\tilde{x}^{(p-1)}]$ is the osculating element of rank p of the curve in $\{\sim A^{h-1}\}$ obtained from $\mathfrak{C}$ by projection along $\{A^{h-1}\}$.

By definition

$$\tilde{m}_p(r;A^h) = \tilde{m}_p(r;A^{h-1}) + \frac{1}{2\pi}\int_0^{2\pi} \log \frac{|[A^{h-1}X^p]|}{|[A^hX^p]|}\cdot d\vartheta,$$

$$\tilde{N}_p(R;A^h) = \tilde{N}_p(R;A^{h-1}) + N(R,\frac{|[A^hX^p]|}{|[A^{h-1}X^p]|}).$$

Therefore the formula (5.5) for h is converted into the corresponding formula for h - 1 by the averaging process $\mathfrak{M}(A^h \supset A^{h-1})$, once the following three facts are established:

(i) $$\mathfrak{M}_{A^h \supset A^{h-1}} \log |[A^h X^p]| = \log |[A^{h-1} X^p]| + \text{const.}$$

(valid for $p + h \leqq n$),

(ii) $$\mathfrak{M}_{A^h \supset A^{h-1}} N\left(R, \frac{|[A^h X^p]|}{|[A^{h-1} X^p]|}\right) = 0 \quad (p + h \leqq n - 1),$$

(iii) $$\mathfrak{M}_{A^h \supset A^{h-1}} S^p(z; A^h) = S^p(z; A^{h-1}) \quad (p + h \leqq n - 1).$$

Indeed, (5.6) combined with (i) and (ii) results in the equation

$$\mathfrak{M}_{A^h \supset A^{h-1}} T_p(R; A^h) = T_p(R; A^{h-1}),$$

which, together with (iii), gets us from h to h - 1 in (5.5).

For h = 1 the formulas (i), (ii), (iii) are equivalent to the following lemmas:

LEMMA 5. A.

$$\mathfrak{M}_a \log |[aX^p]| = \log |X^p| + \text{const.} \quad (p \leqq n - 1).$$

LEMMA 5. B.

$$\mathfrak{M}_a N\left(R, \frac{|[aX^p]|}{|X^p|}\right) = \mathfrak{M}_a \tilde{N}_p(R; a) = 0 \quad (p \leqq n - 2).$$

LEMMA 5. C.

$$\mathfrak{M}_a \frac{|[aX^{p-1}]|^2 \cdot |[aX^{p+1}]|^2}{|[aX^p]|^4} = \frac{|X^{p-1}|^2 \cdot |X^{p+1}|^2}{|X^p|^4} \quad (p \leqq n - 2).$$

However, if we apply these lemmas to the $(n - h + 1)$-space $\{\sim A^{h-1}\}$ instead of the original n-space, we obtain the formulas (i), (ii), (iii) <u>for arbitrary</u> h.

Lemmas A and C have nothing to do with the curve. Here $x, x', \ldots, x^{(p)}$ are any given vectors and

$$X^{p-1} = [x\ldots x^{(p-2)}],\ X^p = [x\ldots x^{(p-1)}],\ X^{p+1} = [x\ldots x^{(p)}].$$

The two propositions are proved by the same trick that served to prove Lemma 2. A. Choose a unitary basis such that (5.4) holds. Then

$$|[aX^p]|^2 = |X^p|^2\cdot(|a_{p+1}|^2 + \ldots + |a_n|^2),$$

and Lemma A follows at once. The constant at the right-hand side (depending on nothing but n and p) has the value

$$\tfrac{1}{2}\mathfrak{M}_a \log(|a_{p+1}|^2 + \ldots + |a_n|^2).$$

The left side of the formula of Lemma C becomes

$$c_p^n \frac{|X^{p-1}|^2\cdot|X^{p+1}|^2}{|X^p|^4}$$

with

$$(5.7)\quad c_p^n = \mathfrak{M}_a \frac{(|a_p|^2+\ldots+|a_n|^2)\cdot(|a_{p+2}|^2+\ldots+|a_n|^2)}{(|a_{p+1}|^2+\ldots+|a_n|^2)^2}\ .$$

No great harm would be done if c_p^n were not unity: we should have to add certain numerical factors to the right members of (5.5). But as a matter of fact we shall presently verify the equation $c_p^n = 1$.

Lemma B is of different character. Remember that all components of X^p have a common zero of order $d_p = d_p(z_0)$ at $z = z_0$. After its removal

$$X^p = (z-z_0)^{d_p}\cdot X^p_{red},$$

the product $[a, X^p_{red}]$ vanishes there to the order

$\nu_p(z_0:a)$, and $\tilde{N}_p(R;a)$ is the corresponding valence

$$\tilde{N}_p(R;a) = \sum_z \phi(R;z)\nu_p(z:a).$$

Adjust the coordinate system to the point $z = z_0$ of the curve, so that the corresponding branch is represented by the expansions

$$x_1 = (z-z_0)^{\delta_1}+\ldots,\ x_2 = (z-z_0)^{\delta_2}+\ldots,\ \ldots,\ x_n = (z-z_0)^{\delta_n}+\ldots$$

$$(0 = \delta_1 < \delta_2 < \ldots < \delta_n).$$

(It must not be overlooked that in general the adjusted coordinates do not form a unitary system!) Writing out the ordered components $(1,\ldots,p,i)$ of the equation $[a,X^p_{red}] = 0$ for $z = z_0 (i = p+1,\ldots,n)$, one finds

$$a_{p+1} = \ldots = a_n = 0,$$

while for z sufficiently near to z_0 relations

$$a_i = \Phi_{i1}(z)a_1 + \ldots + \Phi_{ip}(z)a_p \quad (i=p+1,\ldots,n)$$

result with coefficients Φ regular and vanishing at $z = z_0$. This reveals that $\nu_p(z:a) = 0$ for all z in the circle $|z| \leqq R$ except on an analytic manifold in the a-space of $p + 1$ complex dimensions. Provided $p + 1 \leqq n - 1$ we have therefore

$$\mathfrak{M}_a\tilde{N}_p(R;a) = \mathfrak{M}_a\sum_z \phi(R;z)\nu_p(z:a) = 0,$$

in agreement with Lemma 5. B. Only if $p + 1 = n$ the average must be expected to be positive, and this computation, of a definitely more difficult character than our present argument, has been carried out in §4.

It remains to compute (5.7) and to prove that the integral over the whole a-space of

$$e^{-|a|^2}\cdot\frac{(|a_p|^2+\ldots+|a_n|^2)(|a_{p+2}|^2+\ldots+|a_n|^2)}{(|a_{p+1}|^2+\ldots+|a_n|^2)^2}$$

equals π^n. By the use of polar coordinates in familiar fashion, $|a_k|^2 = t_k$, and by singling out the two indices $p+1$, p: $t_{p+1} = t$, $t_p = s$, we reduce our problem to the equation

$$\int_0^\infty \int_0^\infty \frac{A(A+t+s)}{(A+t)^2} e^{-t-s} dt\, ds = 1$$

where $A = |a_{p+2}|^2 + \dots + |a_n|^2$. The left side equals

$$(5.8) \qquad \int_0^\infty \frac{A}{A+t} \cdot e^{-t} dt + \int_0^\infty \frac{A}{(A+t)^2} e^{-t} dt.$$

Transform the first part by partial integration,

$$-\int_0^\infty \frac{A}{A+t}\, de^{-t} = 1 - \int_0^\infty \frac{A}{(A+t)^2} e^{-t} dt$$

and thus find the value 1 for the sum (5.8).

We have now established by induction the string of equations (5.5) for $h = n - p - 1, \dots, 0$, ending with (5.1). This proof is longer than the first, but gives a deeper insight into the mechanism. The whole Chapter V will be based on a generalization of this argument.

§6. The compensating term for the stationary points

The same expression $S^p = S^p_z$, whose zeros are the stationary points of rank p, occurs in the formulas (5.1), (1.3) for T_p and the quantity Ω_p,

$$2\Omega_p = \frac{1}{2\pi} \int \log \tfrac{1}{2} S^p_z \cdot d\vartheta.$$

This makes it possible to appraise Ω_p in terms of T_p.

To any analytic substitution $z = z(\zeta)$ of the variable z the quantity S^p_z reacts as a density,

$$S^p_z \left| \frac{dz}{d\zeta} \right|^2 = S^p_\zeta,$$

where S^p_ζ is obtained from

$$X^p_\zeta = \left[x, \frac{dx}{d\zeta}, \ldots, \frac{d^{p-1}x}{d\zeta^{p-1}}\right]$$

as S^p_z is from $X^p = X^p_z$. In particular, for the substitution

$$z = r_0 \cdot e^\zeta \qquad (\zeta = u + i\vartheta, \quad u = \log \frac{r}{r_0})$$

we find

$$S^p_z = S^p_\zeta / r^2, \quad \log \tfrac{1}{2} S^p_z = \log S^p_\zeta + \log \tfrac{1}{2} r_0^{-2} - 2u.$$

Along a circle $r =$ const. differentiation with respect to ζ, $\frac{d}{d\zeta}$, equals $\frac{1}{i}\frac{d}{d\vartheta}$. We modify our definition of Ω_p by setting from now on

$$2\Omega_p = \frac{1}{2\pi} \int \log S^p_\vartheta \cdot d\vartheta \tag{6.1}$$

where

$$X^p_\vartheta = \left| x, \frac{dx}{d\vartheta}, \ldots, \frac{d^{p-1}x}{d\vartheta^{p-1}} \right| \quad \text{for } z = re^{i\vartheta},\ r = \text{const.},$$

$$S^p_\vartheta = 2 \cdot \frac{|X^{p-1}_\vartheta|^2 \cdot |X^{p+1}_\vartheta|^2}{|X^p_\vartheta|^4}.$$

Then the Second Main Theorem assumes the form

$$V_p + (T_{p+1} - 2T_p + T_{p-1}) = [\Omega_p]^R_{r_0} - \log \frac{R}{r_0}.$$

In (5.1) we split off the integral over the nucleus while outside we replace

$$S^p_z dz\, d\bar{z} \text{ by } S^p_\zeta d\zeta\, d\bar{\zeta} = S^p_\zeta d\vartheta\, du.$$

Introducing the positive constant

$$c_p = \frac{1}{2\pi} \iint_{|z| \leqq r_0} S^p_z dz\, d\bar{z}$$

we thus obtain

(6.2) $$T_p = c_p \log \frac{R}{r_0} + \int_{u=0}^{\infty} \phi Q^p du$$

with

(6.3) $$Q^p = \frac{1}{2\pi} \int S_\vartheta^p d\vartheta.$$

We now introduce $u = \log \frac{r}{r_0}$ instead of the radius r as the independent variable (logarithmic scale) so that T(u) means what has previously been denoted by $T(r_0e^u)$. Similarly for all other functions of the radius. In all quantities defined by integration with respect to ϑ, like $m(r;\alpha)$ or $Q^p(r)$, the function of z under the integral is changed into one of ϑ by the substitution $z = r_0 \cdot e^{u+i\vartheta} (u \geqq 0)$. The potential ϕ assumes the simple expression

$$\phi = 0 \text{ for } u \geqq U, \quad = U - u \text{ for } 0 \leqq u \leqq U, \quad = U \text{ for } u \leqq 0.$$

Thus (6.2) becomes

$$T_p(U) = c_pU + \int_0^U (U-u)Q^p(u)\, du,$$

or if one prefers,

$$T_p(u) = c_pu + \int_0^u (u-u')Q^p(u')\, du'.$$

This integral equation is equivalent to the differential relations:

(6.4) $$T_p = 0, \quad \frac{dT_p}{du} = c_p \quad \text{for } u = 0;$$
$$\frac{d^2T_p}{du^2} = Q^p(u).$$

The formula of the Second Main Theorem now reads

$$V_p + (T_{p+1} - 2T_p + T_{p-1}) = [\Omega_p]_0^U - U.$$

Once we have completely switched over to the logarithmic scale of the radius we may use the old letter r for u. With due apologies for any confusion thus caused, this will be done from now on; hence r designates the radius measured in the logarithmic scale.

Because log is a concave function, the logarithm of the mean of any quantities exceeds the mean of their logarithms. Applying this remark to (6.3) we find

$$\log Q^p \geqq \frac{1}{2\pi}\int \log S^p_\vartheta \cdot d\vartheta = 2\Omega_p, \tag{6.5}$$

and (6.4) and (6.5) imply the fundamental relation

$$\Omega_p(r) \leqq \frac{1}{2}\log \frac{d^2T_p}{dr^2} .$$

Would the standard function T_p itself, rather than its second derivative, appear on the right side, then we could rightly claim that Ω_p is essentially negative and thus attain the goal we set ourselves at the end of §1. But one knows how precarious conclusion from a function on its second derivative is! With what qualifications then, we must ask, is it true that the logarithm of the second derivative is majorized by the logarithm of the function itself? From the outset it is clear that what matters for our purposes is not the equation but the inequality

$$c_pR + \int_0^R (R-r)Q^p(r)dr \leqq T_p(R),$$

which changes by means of (6.5) into

$$c_pR + \int_0^R (R-r)\cdot e^{2\Omega_p(r)}dr \leqq T_p(R). \tag{6.6}$$

It implies the remarkable inequality

$$T_p(R) \geqq c_pR. \tag{6.7}$$

There is, however, another way of realizing that T_p grows at least as fast as R. Write the First Main

Theorem in the form

$$N_p(R;A) + m_p(R;A) = T_p(R) + m_p^0(A).$$

Choose a point z_0 inside the nucleus $|z| \leqq r_0$, adjust the coordinates to the branch of our curve at z_0, see (I, 7.1) and take an $A = A_0$ whose component $A_0(1,\ldots,p)$ vanishes. Then $\nu_p(z_0;A_0) \geqq 1$ and thus $N_p(R;A_0) \geqq R$. Denoting the constant $m_p^0(A_0)$ by a, we get

$$T_p(R) \geqq R - a. \tag{6.8}$$

Hence

$$1 + (1-c_p)R \leqq T_p(R) \quad \text{as soon as} \quad R \geqq R_0 = \frac{1+a}{c_p}.$$

We add this to (6.6):

$$1 + R + \int_0^R (R-r)\cdot e^{2\Omega_p(r)}\,dr \leqq 2T_p(R) \text{ (for } R \geqq R_0). \tag{6.9}$$

(6.6) has the advantage of holding for all R, whereas (6.9) holds only from a certain $R = R_0$ on (depending on the curve $\mathfrak{C}$). But (6.9) has the advantage that the constant c_p in (6.6) which also depends on the curve. is replaced by 1. The factor 2 in the right member of (6.9) will hurt nobody. We shall presently see why we care to have the additional term 1 in the left member of (6.9). The double aspect represented by (6.6) and (6.9) will be encountered in all our dealings with this situation, but the latter will prove the more important. In any case these considerations show that the combination of the two terms on the left side of (6.6) is of a somewhat artificial character.

The question raised above is answered by the following

LEMMA 6. A. Let two functions $f(r)$, $F(r)$ $(r \geqq 0)$ be given such that

$$1 + r + \int_0^r (r-\rho)e^{f(\rho)}\,d\rho \leqq F(r) \qquad \text{(for } r>0). \tag{6.10}$$

Choose numbers $\kappa > 1$ and d. Then

$$f(r) \leqq \kappa^2 \log F(r) + d \tag{6.11}$$

for all $r > 0$, except in an open set of measure less than

$$M = \frac{2}{\kappa-1} e^{-d'}, \qquad d' = \frac{d}{1+\kappa}. \tag{6.12}$$

A qualification $r > R_0$ might be added to both hypothesis (6.10) and conclusion.

This lemma is an immediate consequence of

LEMMA 6. B. Let $f(r)$ be a function with continuous non-negative derivative and $f(0) \geqq 1$. Then

$$\frac{df}{dr} \leqq e^{d'} \cdot (f(r))^{\kappa},$$

except in an open set of measure less than $\frac{1}{\kappa-1} e^{-d'}$.

Indeed, if in an interval

$$\frac{df}{dr} > e^{d'} \cdot (f(r))^{\kappa}, \tag{6.13}$$

then the integral over that interval satisfies the relation

$$\int dr < e^{-d'} \int \frac{df}{f^{\kappa}},$$

and thus it follows that the integral $\int dr$ over the finite or infinite number of open intervals in which (6.13) holds is less than

$$e^{-d'} \int_1^{\infty} y^{-\kappa} dy = \frac{1}{\kappa-1} e^{-d'}. \tag{6.14}$$

Returning to Lemma A, we set

$$1 + r + \int_0^r (r-\rho) \cdot e^{f(\rho)} d\rho = g(r),$$

so that $g = 1$, $\frac{dg}{dr} = 1$ for $r = 0$ and

$$\frac{d^2g}{dr^2} = e^{f(r)}.$$

Applying Lemma A first to $\frac{dg}{dr}$ and then to g itself, one gets the inequalities

$$e^{f(r)} = \frac{d^2g}{dr^2} \leqq e^{d'}\left(\frac{dg}{dr}\right)^{\kappa},$$

$$\frac{dg}{dr} \leqq e^{d'}\cdot(g(r))^{\kappa},$$

each holding except in an open set of measure less than (6.14). Hence

$$e^{f(r)} \leqq e^{(1+\kappa)d'}\cdot(g(r))^{\kappa^2},$$

except in an open set of measure less than (6.12). The hypothesis of the lemma $g(r) \leqq F(r)$ results in (6.11).

Lemma A is of immediate application to the inequality (6.9), yielding

$$2\Omega_p(r) \leqq \kappa^2 \log T_p(r) + d \qquad \text{for } r > R_0$$

except in a set of measure less than

$$M = \frac{2}{\kappa - 1}\, e^{-d'} \quad \text{where now } (1+\kappa)d' = d - \kappa^2 \log 2.$$

(This upper bound of the measure does not depend on the curve, whereas R_0 does.)

But let us not get lost in such subtleties, and rather be content with the following weaker statement which is considered an integral part of the Second Main Theorem:

THEOREM. Given any numbers $\kappa > 1$ and d, the inequality

$$(6.15) \qquad 2\Omega_p(r) \leqq \kappa\cdot\log T_p(r) - d$$

holds almost everywhere, i. e. with the exception of an open set of finite measure.

§7. Inequalities for orders and stationarity valences of all ranks

Before drawing simple qualitative consequences from this proposition we make a few obvious remarks about inequalities that hold almost everywhere.

LEMMA 7. A. *If each of the inequalities*

$$f_1(r) \leqq 0, \ \dots \ , f_m(r) \leqq 0$$

holds almost everywhere, then they do so simultaneously.

Proof. The join of m sets of measure less than M_1, ... , M_m is of measure less than $M = M_1 + \dots + M_m$. -- This lemma makes it possible to combine "almost everywhere inequalities" by addition and pass from $f_1 \leqq f_2$, $f_2 \leqq f_3$ to $f_1 \leqq f_3$.

LEMMA 7. B. *Under the hypothesis of Lemma A one can ascertain a sequence $r_1 < r_2 < \dots$, $r_\nu \to \infty$ with $\nu \to \infty$, such that $f_i(r_\nu) \leqq 0$ for $i = 1, \dots , m$ and all ν.* A fortiori

$$\varliminf_{r\to\infty} f_i(r) \leqq 0.$$

Indeed, if the inequalities $f_i(r) \leqq 0$ hold except in an open set of measure less than M, then each interval of length M contains values r for which simultaneously $f_i(r) \leqq 0$, $(i = 1, \dots , m)$.

In order to evaluate the Second Main Theorem,

$$(7.1) \quad V_p + (T_{p+1} - 2T_p + T_{p-1}) = (\Omega_p - \Omega_p^0) - R$$

(argument R) let us treat the system of difference equations

$$(7.2) \qquad T_{p+1} - 2T_p + T_{p-1} = -f_p \qquad (p=1,\dots,n-1)$$

in which $f_1, \dots, f_{n-1}$ are given numbers and $T_0, \dots , T_n$

unknowns. By double summation one obtains the solution of the initial value problem:

$$(7.3) \qquad T_p = -(p-1)T_0 + pT_1 - \sum_{q=1}^{p-1}(p-q)f_q,$$

in particular

$$(7.4) \quad T_n = -(n-1)T_0 + nT_1 - \{(n-1)f_1 + \dots + 1\cdot f_{n-1}\}.$$

The solution of the boundary value problem $T_0 = 0$, $T_n = 0$ is obtained by setting $T_0 = 0$ in (7.3) and determining T_1 from (7.4):

$$nT_1 = \sum_1^{n-1}(n-q)f_q.$$

The result is a formula

$$(7.5) \qquad T_p = \frac{1}{n}\sum_q k(p,q)f_q$$

with the following symmetric "Green's kernel"

$$k(p,q) = \begin{cases} q(n-p) & \text{for } q \leq p, \\ p(n-q) & \text{for } p \leq q. \end{cases}$$

(7.4) shows that the initial values T_0, T_1 must satisfy the equation

$$nT_1 - (n-1)T_0 = (n-1)f_1 + \dots + 1\cdot f_{n-1}$$

if the end value T_n is to vanish. Apply this to T_{p-1}, T_p instead of T_0, T_1: Any two consecutive terms T_{p-1}, T_p of a solution with the end value $T_n = 0$ satisfy the equation

$$(7.6) \quad (n+1-p)T_p - (n-p)T_{p-1} = (n-p)f_p + \dots + 1\cdot f_{n-1}.$$

Because of the symmetry of the problem with respect to the inversion $p \to n-p$ we also find the relation

$$(7.7) \qquad (p+1)T_p - pT_{p+1} = pf_p + \dots + 1f_1$$

for any solution with the initial value $T_0 = 0$. If T_0, T_n both vanish, we can use both sets (7.6), (7.7).

Dropping the negative $-R$ and spoiling all the finer points by the rough estimate $\log x = o(x)$ we draw from (7.1) and (6.15) the conclusion that the inequality

$$\| \qquad V_p + (T_{p+1} + T_{p-1}) < kT_p$$

holds almost everywhere if k is any given number > 2; in particular,

$$(7.8) \quad \| \qquad T_{p+1} < kT_p, \qquad T_{p-1} < kT_p.$$

In the following, inequalities holding almost everywhere are marked by $\|$. Simultaneous use of the relations (7.8) for $p = 1, \ldots, n-1$ is granted by Lemma A. Consequently

$$\| \qquad T_1 < kT_2, \ \ldots \ , T_{p-1} < kT_p;$$

$$(7.9) \quad \| \ T_1 \leqq k^{p-1}T_p, \ T_2 \leqq k^{p-2}T_p, \ \ldots \ , T_p \leqq T_p.$$

Let us now apply (7.7) with the right members

$$(7.10) \qquad f_p = (\Omega_p^0 - \Omega_p) + R + V_p,$$

but throw away the part $R + V_p \geqq 0$:

$$pT_{p+1} - (p+1)T_p \leqq 1(\Omega_1 - \Omega_1^0) + \ldots + p(\Omega_p - \Omega_p^0).$$

Combining the relations

$$\| \ \Omega_1 - \Omega_1^0 < \kappa \log T_1 - d, \ \ldots \ , \Omega_p - \Omega_p^0 < \kappa \log T_p - d$$

with (7.9) whence by proper choice of d

$$\| \quad 1(\Omega_1 - \Omega_1^0) + \ldots + p(\Omega_p - \Omega_p^0) < \kappa \cdot \frac{p(p+1)}{2} \log T_p,$$

we find

$$\| \qquad T_{p+1} - (1 + \tfrac{1}{p})T_p < \kappa \frac{p+1}{2} \log T_p,$$

and, because of the symmetry $p \to n-p$, also

$$\| \qquad T_{p-1} - (1 + \frac{1}{n-p})T_p < \kappa \frac{n+1-p}{2} \log T_p.$$

Utilizing the full expression (7.10) of f_p one concludes by a similar argument from (7.5):

$$\| \quad \frac{1}{2}p(n-p)R + \frac{1}{n}\sum_q k(p,q)V_q < T_p + \kappa \frac{p(n-p)}{2} \log T_p.$$

Although the complete relation has its interest, we simplify by picking out the term $q = p$ in the sum on the left side, and according the logarithm the same rough treatment as before we arrive at the following neat

THEOREM. ϵ being any positive number, the inequalities

$$\| \quad T_{p+1} < (1 + \frac{1}{p} + \epsilon)T_p, \quad T_{p-1} < (1 + \frac{1}{n-p} + \epsilon)T_p,$$

$$\| \qquad V_p < (\frac{n}{p(n-p)} + \epsilon)T_p,$$

$$\| \qquad (T_{p+1}+T_{p-1}) + V_p < (2+\epsilon)T_p \quad (p=1,\ldots,n-1)$$

hold almost everywhere.

In this sense all the T_p are of the same order of magnitude, and the V_p not of higher order than they. <u>Thus the transcendency level of the analytic curve is set by the one quantity</u> $T = T_1$.

Still more roughly, the limits inferior $\tau_p, \tau^*_p, \sigma_p$ of T_{p+1}/T_p, T_{p-1}/T_p, V_p/T_p satisfy the relations

$$\tau_p \leqq 1 + \frac{1}{p}, \quad \tau^*_p \leqq 1 + \frac{1}{n-p}, \quad \sigma_p \leqq \frac{n}{p(n-p)} ;$$

$$\tau_p + \tau^*_p + \sigma_p \leqq 2.$$

§8. A peculiar relation

As we have seen in I, §7 the stationarity index $v_p\{\mathfrak{C}\} - 1$ of rank p of the curve $\mathfrak{C}$ at a point $z = z_0$ is $v_1\{\mathfrak{C}_p\} - 1$ and thus $V_p\{\mathfrak{C}\} = V_1\{\mathfrak{C}_p\}$. The first proof of the fundamental equation (5.1) consisted in verification of a similar behavior for the corresponding compensating Ω:

$$S^p\{\mathfrak{C}\} = S^1\{\mathfrak{C}_p\}, \qquad \Omega_p\{\mathfrak{C}\} = \Omega_1\{\mathfrak{C}_p\}.$$

Moreover, definition or the first of these relations gives $T_p\{\mathfrak{C}\} = T_1\{\mathfrak{C}_p\}$. It is natural to ask whether the higher v_q, S^q of the curve $\mathfrak{C}_p$ in $\binom{n}{p}$-space may be expressed by the same quantities for $\mathfrak{C}$. Closer examination shows that such relations prevail only for $q = 2$ besides $q = 1$.

We go back to the notations of Chap. I §7. The least of the exponents $d(i_1,\ldots,i_p)$ is $d(1,\ldots,p) = d_p$, the next is $d(1,\ldots,p-1, p+1)$, and after that comes either $d(1,\ldots,p-2, p, p+1)$ or $d(1,\ldots,p-2, p-1, p+2)$. Hence the δ_2 and δ_3 of $\mathfrak{C}_p$ are v_p and $\min(v_{p-1}+v_p, v_p+v_{p+1})$ respectively; consequently $v_1\{\mathfrak{C}_p\} = v_p$ and

$$(8.1) \qquad v_2\{\mathfrak{C}_p\} = \min(v_{p-1}, v_{p+1}).$$

We now propose to show that a similar relation prevails for $S^2\{\mathfrak{C}_p\}$,

$$(8.2) \qquad S^2\{\mathfrak{C}_p\} = S^{p-1} + S^{p+1}.$$

Proof. Differentiating $\dot{X}_p$, (5.3), one finds

$$\ddot{X}_p = [x,\ldots,x^{(p-3)},x^{(p-1)},x^{(p)}] + [x,\ldots,x^{(p-2)},x^{(p+1)}]$$

$$= \ddot{X}_1^p + \ddot{X}_2^p.$$

Using an adapted unitary coordinate system in which the components of $x, x', \ldots, x^{(p+1)}$ form a recursion scheme like (5.4), and indicating the components

$$(1,\ldots,p),\ (1,\ldots,p-1,p+1),\ (1,\ldots,p-2,p,p+1),$$
$$(1,\ldots,p-2,p-1,p+2)$$

of p-ads by $\hat{1}$, $\hat{2}$, $\hat{3}$, $\hat{4}$ respectively, we obtain the following values

$$X^p(\hat{1}) = x_1 x_2' \ldots x_p^{(p-1)};$$

$$\dot{X}^p(\hat{1}) = *,\ \dot{X}^p(\hat{2}) = x_1 x_2' \ldots x_{p-1}^{(p-2)} x_{p+1}^{(p)};$$

$$\ddot{X}_1^p(\hat{1}) = *,\ \ddot{X}_1^p(\hat{2}) = *,\ \ddot{X}_1^p(\hat{3}) = x_1 \ldots x_{p-2}^{(p-3)} x_p^{(p-1)} x_{p+1}^{(p)};$$

$$\ddot{X}_2^p(\hat{1}) = *,\ \ddot{X}_2^p(\hat{2}) = *,\ \ddot{X}_2^p(\hat{4}) = x_1 \ldots x_{p-1}^{(p-2)} x_{p+2}^{(p+1)}.$$

All other ordered components vanish. Consequently the only non-vanishing ordered components of $\{X^p \dot{X}^p \ddot{X}^p\}$ are

$$\{X^p \dot{X}^p \ddot{X}^p\}(\hat{1},\hat{2},\hat{3}) = (x_1 \ldots x_{p-2}^{(p-3)})^3 (x_{p-1}^{(p-2)} x_p^{(p-1)} x_{p+1}^{(p)})^2,$$

$$\{X^p \dot{X}^p \ddot{X}^p\}(\hat{1},\hat{2},\hat{4}) = (x_1 \ldots x_{p-1}^{(p-2)})^3 x_p^{(p-1)} x_{p+1}^{(p)} x_{p+2}^{(p+1)} ,$$

and therefore

$$\frac{|\{X^p \dot{X}^p \ddot{X}^p\}|^2 \cdot |X^p|^2}{|\{X^p \dot{X}^p\}|^4} = \frac{|x_p^{(p-1)}|^2}{|x_{p-1}^{(p-2)}|^2} + \frac{|x_{p+2}^{(p+1)}|^2}{|x_{p+1}^{(p)}|^2}$$

$$= \frac{|X^{p-2}|^2 \cdot |X^p|^2}{|X^{p-1}|^4} + \frac{|X^p|^2 \cdot |X^{p+2}|^2}{|X^{p+1}|^4} .$$

By the general law (5.1) the equation (8.2) thus proved implies

$$T_2\{\mathbb{C}_p\} = T_{p-1} + T_{p+1},$$

a strange relation of which we are not aware whether it is known even for rational and algebraic curves. Since

$$\log(S^{p-1} + S^{p+1}) \geqq \log S^{p-1} \text{ and } \geqq \log S^{p+1}$$

the same equation combined with the definition (1.3) gives rise to the inequality

$$\Omega_2\{\mathbb{C}_p\} \geqq \max(\Omega_{p-1}, \Omega_{p+1})$$

which runs parallel to (8.1).

CHAPTER IV.

FIRST AND SECOND MAIN THEOREMS FOR ANALYTIC CURVES

§1. Green's formula

On a given Riemann surface $\mathfrak{R}$ the neighborhood of any point $\mathfrak{p}_0$ may be referred to a local parameter $z = x + iy$. With any function ϕ there is associated the invariant differential

$$(1.1) \qquad d\phi = \frac{\partial\phi}{\partial x}\, dx + \frac{\partial\phi}{\partial y}\, dy$$

provided the function has continuous first derivatives around $\mathfrak{p}_0$ (is of class D_1). Rotation by 90° carries the line element (dx, dy) into one with the components

$$\delta x = -dy, \qquad \delta y = dx;$$

as this process is independent of the choice of the local parameter the linear differential

$$(1.2) \qquad \frac{\partial\phi}{\partial y}\, dx - \frac{\partial\phi}{\partial x}\, dy = d^{\perp}\phi$$

is also invariant with respect to analytic transformation of $z = x + iy$. If the line element (dx, dy) is an element of a contour one often writes $\frac{\partial\phi}{\partial s}\, ds$, $\frac{\partial\phi}{\partial n}\, ds$ for (1.1) and (1.2) [tangential and normal derivatives]. In case ϕ is of class D_2 and harmonic,

$$(1.3) \qquad d\zeta = \left(\frac{\partial\phi}{\partial x} - i\,\frac{\partial\phi}{\partial y}\right) dz = d\phi - i d^{\perp}\phi$$

is an analytic differential, and ϕ has, at least locally, a conjugate ϑ such that $\zeta = \phi - i\vartheta$ is an analytic function of which (1.3) is the differential, $d^{\perp}\phi = d\vartheta$.

Let G be a compact part of the surface bounded by a finite number of Jordan contours Γ no two of which have a point in common, and

$$\psi = \psi_x dx + \psi_y dy$$

an invariant linear differential. The expression

$$\frac{\partial \psi_y}{\partial x} - \frac{\partial \psi_x}{\partial y} = \mathrm{rot}_z \psi = \psi'_z$$

is transformed according to the law

$$\psi'_z \cdot \left| \frac{dz}{dz^*} \right|^2 = \psi'_{z^*},$$

under any transformation $z \to z^*$ of the local parameter. Hence ψ'_z is the expression in terms of z of a certain <u>density</u> ψ'. We wish to establish <u>Stokes's formula</u>

(1.4) $$\int_\Gamma \psi = \int_G \psi',$$

in which the customary convention about the positive sense along Γ has been adopted.

Fortunately we need this formula only for regions G bounded by <u>analytic</u> contours Γ which separate G from its complement $\overline{G}$. More precisely: with each point $\mathfrak{p}_0$ on the contour we can associate a local parameter $z = x + iy$ at $\mathfrak{p}_0$ and a rectangle P: $|x| \leqq a$, $|y| \leqq b$ which is the one-to-one conformal map of a block P on $\mathfrak{R}$ (admissible parameter and block) such that the upper half of P, $y \geqq 0$, represents the intersection $P \cap G$ of the block P with G. For an interior point $\mathfrak{p}_0$ of G any local parameter z and a block P, $|x| \leqq a$, $|y| \leqq b$, which lies entirely inside G are admissible. The salient point is that for each of these admissible blocks P the (x,y)-image of the intersection $P \cap G$ is of such shape (namely rectangular) that the integral of a continuous function of x, y over it may be defined by subdivision into small rectangles without any difficulties arising from the boundary. We may now construct Dieudonné factors μ_q by means of a finite number of admissible blocks covering G. We see to it that $\mu(x,y)$ is of class D_1 in P.

Let Δ be a continuous density and Δ_z, Δ_{z*} its expressions in terms of two local parameters z, z* associated with the points p_0, p_0^*, so that at any point covered by both

$$\Delta_z \cdot \left| \frac{dz}{dz*} \right|^2 = \Delta_{z*}.$$

Suppose moreover that both z, z* with their blocks P, P^* are admissible and that Δ vanishes outside P as well as P^*. Then

$$(1.5) \qquad \int\int_{P \cap G} \Delta_z dx\, dy = \int\int_{P^* \cap G} \Delta_{z*} dx^* \, dy^* .$$

By means of this basic lemma and proceeding in the same manner as in Chap. III, §3, one can define the integral $\int_G \Delta$ over G of any continuous density Δ in G. Thus the right side of Stokes's formula acquires a definite meaning.

Dieudonné's partition $\psi = \sum \mu_q \psi$ reduces Stokes's formula to the corresponding equations for the parts,

$$(1.6) \qquad \int_\Gamma \mu_q \psi = \int_G (\mu_q \psi)' .$$

One sees here why it is necessary to choose the μ's as differentiable functions. After writing

$$\omega = \mu \psi \qquad\qquad (\mu = \mu_q)$$

(1.6) is equivalent to the following relations,

$$(1.7) \qquad \int_{-b}^{b} \int_{-a}^{a} \left(\frac{\partial \omega_y}{\partial x} - \frac{\partial \omega_x}{\partial y}\right) dx\, dy = 0$$

for an inner block,

$$(1.7') \qquad \int_0^b \int_{-a}^{a} \left(\frac{\partial \omega_y}{\partial x} - \frac{\partial \omega_x}{\partial y}\right) dx\, dy = \int_{-a}^{a} \omega_x(x,0)\, dx$$

for a boundary block. In this form they can at once be verified, and it is here where the positive sense on Γ comes into play.

Perhaps the proof becomes clearer if we establish the relation (1.4) on the basis of a definite Dieudonné partition without bothering whether the resulting surface integral is independent of the auxiliary construction. Summation of the formulas (1.7), (1.7') corresponding to the admissible covering blocks P_q with their probability factors μ_q yields for the line integral $\int_\Gamma \psi$ the value

$$\sum_q \iint_{P_q \cap G} (\mu_q \psi)'_{z_q} dx_q dy_q .$$

We write it as a double sum

$$\sum_{q,s} \iint_{P_q \cap G} \mu_s (\mu_q \psi)'_{z_q} dx_q dy_q .$$

The integrand of the individual term (q,s) vanishes both outside P_q and P_s. Hence by the lemma mentioned above that term may be transformed into

$$\iint_{P_s \cap G} \mu_s (\mu_q \psi)'_{z_s} dx_s \, dy_s ,$$

and summation with respect to q then yields the desired relation

$$\int_\Gamma \psi = \sum_s \iint_{P_s \cap G} \mu_s \cdot \psi'_{z_s} dx_s \, dy_s .$$

Let ϕ_1, ϕ be functions in G of class D_1 and D_2 respectively. We apply Stokes's formula to the linear differential $\psi = \phi_1 \cdot d^{\perp}\phi$. If moreover ϕ is harmonic, then a well-known computation gives

$$-\psi'_z = \frac{\partial \phi_1}{\partial x} \cdot \frac{\partial \phi}{\partial x} + \frac{\partial \phi_1}{\partial y} \cdot \frac{\partial \phi}{\partial y}$$

and hence

$$(1.8)\quad -\int_\Gamma \phi_1 \cdot d^{\perp}\phi = -\int_\Gamma \phi_1 \cdot d\vartheta = \sum_q \int\int_{P_q \cap G} \mu_q \left(\frac{\partial \phi_1}{\partial x_q}\frac{\partial \phi}{\partial x_q} + \frac{\partial \phi_1}{\partial y_q}\frac{\partial \phi}{\partial y_q}\right) dx_q \, dy_q$$

If ϕ_1 is likewise harmonic we interchange ϕ and ϕ_1 and by comparison obtain <u>Green's formula</u>

$$(1.9)\qquad \int_\Gamma (\phi_1 d\vartheta - \phi d\vartheta_1) = 0,$$

where $d\vartheta$, $d\vartheta_1$ are defined by the analyticity of $d\zeta = d\phi - i d\vartheta$ and $d\zeta_1 = d\phi_1 - i d\vartheta_1$. The special case $\phi_1 = 1$ ($\vartheta_1 = 0$) is noteworthy:

$$(1.10)\qquad \int_\Gamma d^{\perp}\phi = \int_\Gamma d\vartheta = 0.$$

For any two functions ϕ, ϕ_1 of class D_1 the expression

$$(\text{grad } \phi_1 \cdot \text{grad } \phi) = \frac{\partial \phi_1}{\partial x}\cdot\frac{\partial \phi}{\partial x} + \frac{\partial \phi_1}{\partial y}\cdot\frac{\partial \phi}{\partial y} = \Delta_z$$

behaves like a density with respect to analytic transformations of the parameter $z = x + iy$. Its integral, the right member of (1.8) is the Dirichlet integral $\int_G \Delta = D_G[\phi_1, \phi]$, and thus we may write that formula as follows:

$$-\int_\Gamma \phi_1 d\vartheta = D_G[\phi, \phi_1].$$

In (1.9) both ϕ and ϕ_1 are supposed to be harmonic, in (1.8) only ϕ.

It is not worth our while to generalize Stokes's formula to contours of more general character. Just as contours in a plane are approximated by polygons, so one may approximate contours on a Riemann surface by piecewise analytic lines. The operation Λ of passing to the limit will not be carried out in Stokes's equation itself but in certain relations derived from it by processes Π the legitimacy of which after passage to the limit would become dubious. In other words, the correct order of operations is: first Π then Λ, and not the other way round.

§2. The condenser formula

From now on we assume the given Riemann surface $\mathfrak{R}$ to be non-compact. A "Hohlraum" G on $\mathfrak{R}$ which contains a conductor K_0 is used as condenser. We keep K_0 fixed, but let G grow so as to exhaust the whole Riemann surface. The potential Φ of the condenser is a continuous function on $\mathfrak{R}$ which is 0 outside G, 1 in K_0 and harmonic in the vacuum $H = G - K_0$. The inner layer γ, the boundary of K_0, will then carry a certain positive charge c, the "capacity" of the condenser, and the outer layer, the boundary of G, will carry the opposite charge -c. The function $R \cdot \Phi(\mathfrak{p})$ is the solution of the electrostatic problem for a given potential difference R between inner and outer layers, and the charge e will be connected with R by the relation $e = c \cdot R$. We shall fix R so, $R = 1/c$, $\phi(\mathfrak{p}) = R \cdot \Phi(\mathfrak{p})$, as to make $e = 1$. We call R (the reciprocal capacity) the voltage of the condenser G.

Before proceeding further let us fix the conditions more precisely. The bar, $\overline{G}$, will be used to indicate the complement of any part G of $\mathfrak{R}$. An open set is said to be limited if its closure is compact. Let K_0 and $\overline{G}$ be closed sets without common points, K_0 compact and G limited. We assume that $\overline{K}_0$ has no limited component. In the process of exhaustion G would finally fill up such a hole, Φ would assume the constant value 1 in it, and it would then be reasonable to let K_0 swallow the hole. For similar reasons we do not allow G to have a component which is free from points of K_0. Finally we assume that K_0 and G are bounded by a finite number of disjoint analytic Jordan contours $\gamma = \sum \gamma_i$, $\Gamma = \sum \Gamma_k$. According to the maximum principle the harmonic function Φ in the vacuum $H = G - K_0$ which has the boundary values 1 on γ and 0 on Γ will satisfy the inequality $0 \leq \Phi \leq 1$ throughout H.

For a point $\mathfrak{p}_0$ on Γ we use an "admissible" local parameter $z = x + iy$ which maps a piece of the contour onto a piece of the real axis $y = 0$ with G lying on the positive side $y \geqq 0$. By Schwarz's principle of symmetry the analytic function $F = \Phi - i\Theta$ may be continued beyond the contour. Consequently the derivatives exist even along the boundary $y = 0$ and

$$\frac{\partial \Phi}{\partial x} = 0, \quad \frac{\partial \Phi}{\partial y} \geqq 0 \qquad \text{for } y = 0.$$

Because $\Phi \geqq 0$ for $y \geqq 0$, the well-known picture of the real part of an analytic function in the neighborhood of a critical point $\mathfrak{p}_0$ where its derivative vanishes proves that our point $\mathfrak{p}_0$ cannot be critical and that the inner normal derivative $\frac{\partial \Phi}{\partial n}$ is actually positive -- unless Φ vanishes identically around $\mathfrak{p}_0$ and thus throughout a certain component of $G - K_0$. Because $\Phi = 1$ in K_0 this component cannot border on K_0, which contradicts our assumption that there is no component of G free from points of K_0. Hence $\frac{\partial \Phi}{\partial n}$ is actually positive on Γ, the density of charge $-\frac{1}{2\pi}\cdot\frac{\partial \Phi}{\partial n}$ is negative, and we find a negative total charge $-c$ on the outer layer given by

$$2\pi c = \int_\Gamma d^\perp \Phi = \int_\Gamma d\Theta.$$

Here Γ is described in such a sense that G lies to the left. Let the normal n on γ point toward the interior of the vacuum. For the same reasons $-\frac{1}{2\pi}\cdot\frac{\partial \Phi}{\partial n}$, the density of charge on γ, is positive; and its total charge $\frac{1}{2\pi}\int_\gamma d\Theta$ must have the value c, according to the equation (1.10) applied to H. Here γ is described in such a sense that K_0 (not H) lies to the left. Forming $\phi = \frac{1}{c}\Phi$ we have

$$\int_\gamma d\vartheta = \int_\Gamma d\vartheta = 2\pi, \qquad \phi = R = 1/c \qquad \text{in } K_0,$$

and $d\vartheta$ is positive along γ and Γ. There is only a finite number of <u>critical points</u> in H where the regular analytic differential

$$d\zeta = \left(\frac{\partial \phi}{\partial x} - i\,\frac{\partial \phi}{\partial y}\right) dz$$

vanishes.

If one replaces G by a larger region $G' \supset G$, then $\Phi(\mathfrak{p})$ will increase everywhere. Indeed, the harmonic function

$$\Phi' - \Phi \text{ in } H \qquad (\Phi' = \Phi_{G'}, \; \Phi = \Phi_G)$$

vanishes along γ and is non-negative along the boundary Γ of G, hence non-negative throughout H. It vanishes in K_0 and is non-negative in $\overline{G}$; hence

$$\Phi'(\mathfrak{p}) \geqq \Phi(\mathfrak{p}) \tag{2.1}$$

everywhere. Consequently

$$-\frac{\partial \Phi'}{\partial n} \leqq -\frac{\partial \Phi}{\partial n} \text{ along } \gamma, \tag{2.2}$$

i. e. the density along γ decreases and so does the total charge, $c' \leqq c$ or

$$R' \geqq R. \tag{2.3}$$

(2.1), (2.3) imply $\phi'(\mathfrak{p}) \geqq \phi(\mathfrak{p})$ but it is better to stick to the sharper inequality (2.1), namely

$$\phi(\mathfrak{p})/R \leqq \phi(\mathfrak{p})/R'. \tag{2.4}$$

Instead of (2.2) we may write

$$d\vartheta'/R' \leqq d\vartheta/R \quad \text{or} \quad \frac{d\vartheta'}{d\vartheta} \leqq \frac{R'}{R} \quad \text{on } \gamma. \tag{2.5}$$

Let f be a given meromorphic function on $\mathfrak{R}$. Following J. Weyl*, apply Green's formula (1.9) to the region H, the solution ϕ of the electrostatic problem and the harmonic function $\phi_1 = \log|f|$. The latter function has the zeros and poles of f for singularities. With each such point $\mathfrak{p}_0$ we associate a local parameter z and cut it out by a little circular neighborhood k_ϵ, $|z| < \epsilon$. For the

*Ann. of Math. 42, 1941, 371-408.

moment we assume that no zero nor pole of f lies on the boundaries γ, Γ. We travel along γ, Γ and the circumference $k_\epsilon^!$ of k_ϵ leaving to the left K_0, G, k_ϵ respectively. By the way, at each boundary point $\mathfrak{p}_0$ we may use $\zeta - \zeta_0$, $\zeta = \phi - i\vartheta$, as the local parameter in terms of which a piece of the boundary is represented by the real axis $\Im\zeta = 0$. Green's formula now reads

$$\int_\Gamma(\phi_1 d\vartheta - \phi d\vartheta_1) - \int_\gamma(\phi_1 d\vartheta - \phi d\vartheta_1) = \sum\int_{k_\epsilon^!}(\phi_1 d\vartheta - \phi d\vartheta_1).$$

With ϵ tending to zero ϕ_1 becomes infinite as const. $\log 1/\epsilon$ while $\int_{k_\epsilon^!}|d\vartheta|$ is $O(\epsilon)$. Let $\nu = \nu(\mathfrak{p}_0)$ be the order of f at the center $\mathfrak{p}_0$ of the circle. Then $-\int_{k_\epsilon^!}d\vartheta_1$ equals $2\pi\nu$, and hence the individual term on the right tends to $2\pi\nu(\mathfrak{p}_0)\phi(\mathfrak{p}_0)$. We add the corresponding equation for K_0 with ϕ = const. = R in K_0 and hence $d\vartheta = 0$ along the inner bank of γ. Whereas $d\vartheta$ has different values on the inner and outer banks of γ, the differential $d\vartheta_1$ is the same along both. Hence the result is the following <u>condenser formula</u>

$$(2.6) \quad \frac{1}{2\pi}\int_\Gamma \log|f|\cdot d\vartheta - \frac{1}{2\pi}\int_\gamma \log|f|\cdot d\vartheta = \sum \nu(\mathfrak{p})\phi(\mathfrak{p}),$$

with the sum extending over the entire Riemann surface $\mathfrak{R}$.

The analogy to the corresponding principle in the theory of meromorphic functions in the z-plane is obvious: the circle of radius R has been replaced by the arbitrary region G, and $d\vartheta$ takes the place of the differential of the argument ϑ of z.

The proof of Green's formula as given in §1 is applicable to the region H punctured by the little circular holes; but in carrying it through one has to cover the perimeters of these little circles by admissible blocks. It seems more natural to use the following slightly modified procedure for generalizing Stokes's formula to the case where the differential has a finite number of singular points $\mathfrak{p}_1$, $\mathfrak{p}_2$, ... in G. We see to it that the blocks assigned to points $\neq \mathfrak{p}_i$ never cover $\mathfrak{p}_i$. With this precaution we cover G by a finite number of admissible blocks P_q; those assigned to the singular points will of necessity be among them. We cut out the singular point $\mathfrak{p}_1$, which is the center of its rectangle P_1, by a rectangular "window" so small that no other of the blocks P_q penetrates into it, and extend integration over the perforated instead of the complete P_1. The salient point is that for the purpose of calculating Riemann integrals over this space one can divide it into a regular pattern of arbitrarily small rectangles. This modification of our procedure becomes imperative if a singular

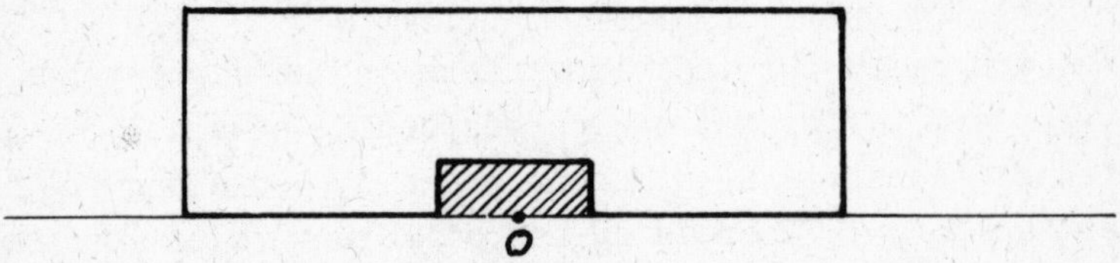

FIG. 2

point $\mathfrak{p}_0$ lies on the boundary Γ. Again we cut a little rectangular indentation over which integration does not extend into the half rectangle $P \cap G$ as indicated by figure 2, and then let the indentation close in on $\mathfrak{p}_0$. In this way we establish the condenser formula for the case where f has zeros or poles on γ or Γ; the line integrals are then to be interpreted as improper integrals.

§3. The first main theorem and the valence integral for T

Armed with the condenser formula we encounter no essential difficulty in carrying over the whole theory of meromorphic curves as far as it centers around the order function. Let $\mathfrak{C}$ be a non-degenerate analytic curve in n-space of type $\mathfrak{R}$ with coordinates x_i. The quotient of two linear forms $(\alpha x) = \sum \alpha_i x_i$ is a meromorphic function f on $\mathfrak{R}$ and application of the condenser formula to this function results in the

FIRST MAIN THEOREM. Let

$$N(\alpha) = \sum_{p} \nu(p;\alpha)\phi(p)$$

be the valence in G of the points of intersection of the curve $\mathfrak{C}$ with the plane (α) and set

$$m(\alpha) = \frac{1}{2\pi}\int_{\Gamma} \log \frac{1}{\|\alpha x\|} \cdot d\vartheta, \quad m^0(\alpha) = \frac{1}{2\pi}\int_{\gamma} \log \frac{1}{\|\alpha x\|} \cdot d\vartheta.$$

Then

$$(3.1) \qquad N(\alpha) + \{m(\alpha) - m^0(\alpha)\} = T$$

is independent of the intersecting plane (α).

If we want to emphasize dependence on G we write $\phi_G(p)$; $N[G;\alpha]$, $m[G;\alpha]$, $m^0[G;\alpha]$, $T[G]$ instead of $\phi(p)$; $N(\alpha)$, $m(\alpha)$, $m^0(\alpha)$, T. Notice that even $m^0(\alpha)$ will in general depend on G because the distribution $d\vartheta$ of electricity on γ does. Since $d\vartheta > 0$ on γ and on Γ both quantities $m(\alpha)$ and $m^0(\alpha)$ are non-negative.

FIRST MAIN THEOREM, continued. A formula similar to (3.1) holds for every rank p,

$$(3.2) \qquad N_p(A) + m_p(A) - m_p^0(A) = T_p$$

in which $\{A\}$ designates an arbitrary contravariant p-element.

The considerations concerning multiplication and projection of analytic curves carry over without modifications.

Next we average with respect to α over the unit sphere. Since $d\varsigma$ is analytic on the analytic contours γ and Γ, the lemmas of II, §4, are applicable, and we find that both averages $\mathfrak{M}m(\alpha)$ and $\mathfrak{M}m^0(\alpha)$ have the same value m_0, and thus (3.1) gives

$$(3.3) \qquad T = \mathfrak{M}N(\alpha).$$

Computation of the average leads to the integral formula

$$(3.4) \qquad T = \frac{1}{2\pi}\int \phi S$$

where S is the density defined by

$$(3.5) \qquad S_z = 2\,\frac{|[xx']|^2}{|x|^4} \qquad \left(x_i' = \frac{dx_i}{dz}\right)$$

in terms of any local parameter z.

Because $\phi_G(\mathfrak{p})/R$ increases with G the same is true for the quotients

$$N[G;\alpha]/R[G] \quad \text{and} \quad T[G]/R[G].$$

By means of the osculating element

$$X_z^p = [x, \frac{dx}{dz}, \ldots , \frac{d^{p-1}x}{dz^{p-1}}]$$

we form the density

$$(3.6) \qquad S_z^p = 2\,\frac{|X_z^{p-1}|^2 \cdot |X_z^{p+1}|^2}{|X_z^p|^4}\,.$$

Its zeros are the stationary points of rank p.

THEOREM. The order T_p of rank p is given by the integral

$$(3.7) \qquad T_p = \frac{1}{2\pi}\int \phi S^p.$$

Let $e - 1 = e(\mathfrak{p}_0) - 1$ denote the order of the regular analytic differential $d\zeta$ at any point $\mathfrak{p}_0$ in H. The critical points $\mathfrak{w}$ are those where this order is positive. A local parameter z may be introduced by $\zeta - \zeta_0 = z^e$. The level lines ϕ = const. of the potential ϕ in H are analytic curves, with one slight exception: the level line passing through a critical point of multiplicity $e - 1$ displays in the neighborhood of that point the rosette characterized by the equation $\Re z^e = 0$. For any value r in the interval $0 < r \leqq R$ we define a region G_r by the inequality $\phi > R - r$. It encloses the nucleus K_0, is part of G and is bounded by the level line Γ_r, $\phi = R - r$. The potential $\phi(r,\mathfrak{p})$ of the condenser G_r with the fixed inner conductor K_0 is given by the equation

$$\phi(r;\mathfrak{p}) = \begin{cases} \phi(\mathfrak{p}) - (R-r) & \text{where } \phi(\mathfrak{p}) \geqq R - r, \\ 0 & \text{where } \phi(\mathfrak{p}) \leqq R - r. \end{cases}$$

Indeed, the normal derivative of $\phi(r;\mathfrak{p})$ along γ coincides with that of $\phi(\mathfrak{p})$ and hence the charge remains unity. For a given point $\mathfrak{p}$, $\phi(r;\mathfrak{p})$ is the following elementary non-negative, non-decreasing convex function of r:

$$0 \text{ for } r \leqq R - \phi(\mathfrak{p}), \qquad r - (R - \phi(\mathfrak{p})) \text{ for } r \geqq R - \phi(\mathfrak{p}).$$

The value of $\phi(r;\mathfrak{p})$ in K_0 is r; hence r is the voltage of the condenser $H_r = G_r - K_0$. Traveling along Γ_r so that G_r is at the left, we have $d\vartheta > 0$ throughout (except at a critical point) and

$$\int_{\Gamma_r} d\vartheta = \int_{\gamma} d\vartheta = 2\pi.$$

We denote by $N(r;\alpha)$ and $T(r)$ the quantities $N[G;\alpha]$ and $T[G]$ for $G = G_r$, in particular

$$N(r;\alpha) = \sum \nu(\mathfrak{p};\alpha)\phi(r;\mathfrak{p})$$

and set

$$m(r;\alpha) = \frac{1}{2\pi}\int_{\Gamma_r} \log \frac{1}{\|\alpha x\|}\cdot d\vartheta (\geqq 0).$$

Then

$$(3.8) \qquad N(r;\alpha) + m(r;\alpha) - m^0(\alpha) = T(r)$$

is independent of (α), and

$$T(r) = \frac{1}{2\pi}\int \phi(r;\mathfrak{p})S(\mathfrak{p}).$$

(Notice that although $m^0[G;\alpha]$ depends on G, $m^0[G_r;\alpha] = m^0[G;\alpha]$ is independent of r.) Because of the properties of the elementary function $\phi(r;\mathfrak{p})$ of r, $N(r;\alpha)$ and $T(r)$ are likewise of regular type, i. e. non-negative, non-decreasing, and convex. The corresponding formulas for rank p offer no difficulties.

At a critical point $\mathfrak{p}_0$ the condition of analyticity is violated for that level line Γ_r which passes through $\mathfrak{p}_0$. In the neighborhood of $\mathfrak{p}_0$ we use $\zeta - \zeta_0$ as the independent variable in terms of which double integrals are expressed. The e tips of G_r grouped around $\mathfrak{p}_0$ are each, when properly sliced off, represented by a rectangle

$$0 < \Re(\zeta-\zeta_0) \leqq a, \quad |\Im(\zeta-\zeta_0)| \leqq b$$

on the positive side of the line $\Re(\zeta-\zeta_0) = 0$. As $\zeta - \zeta_0$ is not a local parameter at $\mathfrak{p}_0$ we cut a dent into the rectangle around ζ_0 as indicated by Figure 2 and then let the indentation close in on ζ_0. (This is all the more required if f happens to have a zero or pole at the critical point $\mathfrak{p}_0$.) Either in this way or by continuity we realize that our formulas hold also for the "critical values" of r for which the contour Γ_r passes through a critical point.

Nothing prevents us from using $\zeta - \zeta_0$ as the independent variable in the neighborhood of any point $\mathfrak{p}_0$ of H, critical or not. Then the part of the integral $\int \phi S^p$ extending over H assumes the form

$$\iint_H \phi S_\zeta^p d\vartheta\, d\phi. \tag{3.9}$$

The integration with respect to ϑ runs along the level line $\phi(\mathfrak{p}) = \phi = \text{const.}$ where $\frac{d}{d\zeta}$ equals $i\,\frac{d}{d\vartheta}$. We therefore write

$$X^p_\vartheta = [x, \frac{dx}{d\vartheta}, \ldots, \frac{d^{p-1}x}{d\vartheta^{p-1}}] \quad \text{for } \phi = \text{const} = R - r,$$

$$S^p_\vartheta = 2\,\frac{|X^{p-1}_\vartheta|^2 \cdot |X^{p+1}_\vartheta|^2}{|X^p_\vartheta|^4},$$

$$2\pi Q_p(r) = \int_{\Gamma_r} S^p_\vartheta d\vartheta.$$

Then (3.9) changes into

$$2\pi \int_0^R (R-r)Q_p(r)dr.$$

Adding the integral of ϕS^p over K_0 which equals R times

$$2\pi c_p = \int_{K_0} S^p$$

we find

$$T_p = c_p R + \int_0^R (R-r)Q_p(r)dr. \tag{3.10}$$

Notice that the positive constant c_p is independent of G. Application of the resulting formula to G_r instead of G gives

$$T_p(r) = c_p r + \int_0^r (r-\rho)Q_p(\rho)d\rho.$$

The quantity r corresponds to the logarithm of the radius in the z-plane (Chaps. II and III), which, from Chap. III, §6 on, has been designated by the same letter.

§4. Existence of the condenser potential and relaxation of conditions for boundary

At this juncture it seems appropriate to describe in broad outline how the potential Φ of the condenser $G - K_0$ may be constructed. We open a competition to all functions Ψ of class D_1 in the closure H' of H which assume the values 1, 0 on γ and Γ respectively; that function Ψ will win for which the energy $\int_H (\text{grad}\,\Psi)^2$ assumes minimum value (Dirichlet's principle). In order to compute the energy integral we cover $H^{\cdot}$ by admissible blocks P_q as described in §1.

Consider one such rectangle P associated with a boundary point $\mathfrak{p}_0$ on Γ, in the plane of the corresponding local parameter $z = x + iy$ and extend the function Ψ, defined in the upper half $y \geqq 0$ of P to the entire rectangle by the condition of "oddity"

$$\Psi(\bar{z}) = -\Psi(z). \tag{4.1}$$

Because Ψ vanishes on Γ, the function thus continued beyond the boundary is still of class D_1. The method described in the author's "Die Idee der Riemannschen Fläche", pp. 104-106, for the construction of a minimizing function, uses the device of replacing an admissible function Ψ inside a circle by the harmonic function Ψ^* which coincides with Ψ along the periphery. With the "odd" continuation this method works also for points on the boundary (inside P). The harmonized Ψ^* will be odd, $\Psi^*(\bar{z}) = -\Psi^*(z)$, because Ψ is. Hence the result is a harmonic function Φ which can be extended as an odd harmonic function a little beyond the boundary. More precisely, this continuation beyond the open segment of the real axis contained in P_q takes place in each of the boundary blocks P_q after Φ has been expressed in terms of the corresponding z_q. (That no contradiction results along those parts of the boundary where several P_q overlap is understandable on the ground of Schwarz's symmetry principle. But since the construction itself furnishes the boundary condition $\Phi = 0$ on Γ, in this stronger form, an explicit appeal to that principle, as in §2, becomes superfluous.) On γ the symmetry condition (4.1) is to be replaced by $\Phi(\bar{z}) = 2 - \Phi(z)$.

The construction could not start unless we were in possession of at least one admissible function Ψ. This point can easily be settled as follows. We assign to each point $\mathfrak{p}_0$ of K_0 a block $P(\mathfrak{p}_0)$ of center $\mathfrak{p}_0$ which does not penetrate into $\bar{G}$, choose a fixed positive number $\theta <$ 1, and select a finite number of points $\mathfrak{p}_1, \mathfrak{p}_2, \ldots$ in K_0 such that the corresponding contracted blocks $P^\theta(\mathfrak{p}_q)$ cover K_0. With a basic probability function $\lambda(x)$ of class D_1 we construct the Dieudonné factors $\mu_q (q=1,2,\ldots)$ for this covering, as described in Chap. III, §3. Then

$$\Psi = \sum \mu_q \ (= \lambda_1 \vee \lambda_2 \vee \ldots)$$

equals 1 in K_0, 0 in $\bar{G}$, and is of class D_1 everywhere.

We could be satisfied with solving the electrostatic problem for analytic contours, were it sure that the Riemann surface is exhaustible by regions G thus bounded. This is certainly feasible with piecewise analytic contours. For a first orientation assume that G in the neighborhood of a point $\mathfrak{p}_0$ on Γ is represented in terms of a suitable local parameter z by a sector $0 < \arg z < \pi\alpha$ of a small circle around z = 0 ("analytic corner"). By $z = \mathfrak{z}^\alpha (\mathfrak{z} = \mathfrak{x} + i\mathfrak{y})$ we map that sector upon a half circle $\mathfrak{y} > 0$ and submit the competing functions Ψ to the condition that they have continuous first derivatives with respect to $\mathfrak{x}, \mathfrak{y}$ even on the bounding diameter $\mathfrak{y} = 0$ of the half circle, including the center $\mathfrak{z} = 0$. Then we can continue into the lower half by the condition $\Psi(\bar{\mathfrak{z}}) = -\Psi(\mathfrak{z})$, although the lower half is not the map of anything on the Riemann surface and is in this sense fictitious. We shall therefore obtain a solution Φ of the electrostatic problem which, when expressed in terms of $\mathfrak{z}$ in the neighborhood of the analytic corner $\mathfrak{p}_0$ and continued into the lower half circle by the symmetry condition $\Phi(\bar{\mathfrak{z}}) = -\Phi(\mathfrak{z})$, is harmonic in the full $\mathfrak{z}$-circle.

When for the investigation of the neighborhood of a critical point on the boundary Γ_r of G_r we made use of the independent variable $\zeta - \zeta_0$, we already abandoned local parameters for the complete neighborhood of $\mathfrak{p}_0$ in favor of what may be termed a straightening parameter, a parameter for the inside half of the neighborhood which maps the boundary onto a straight segment. Once having conceived this idea, one realizes that it works for any *Jordan contour*. A Jordan contour Γ on $\mathfrak{R}$ does not necessarily divide $\mathfrak{R}$ into two parts, but it has two *banks*. The exact formulation of this fact is given by the following construction which takes place in a conformal z-neighborhood $\mathfrak{N}$ of a point O on Γ (see Appendix at end of chapter).

One can draw two cross cuts λ = AB and λ' = A'B' which have no points in common with Γ except their end points A,B and A',B' respectively. These lie on Γ, A and A' on one side, B and B' on the other side of O. A "path" in the z-plane is a finite sequence of oriented segments in which the starting point of one segment coincides with the end point of the preceding one while the segments have no other common points. λ and λ' may be assumed to be paths of this elementary nature. In the z-plane the arc BOA of Γ together with λ forms a Jordan contour the interior of which (λ) lies in $\mathfrak{N}$; similarly λ' and (λ'). *The cross cuts λ and λ' are such that (λ) and (λ') are disjoint and every point in a sufficiently small neighborhood of O lies either in (λ) or in (λ') or on Γ.* If G is bounded by a finite number of disjoint Jordan contours and $\mathfrak{N}$ is free from points on the other contours, then (λ) or (λ') or both will be part of G. (It could happen indeed that they both belong to G, in which case Γ does not separate G from $\overline{G}$; this will necessarily be so if Γ is a Jordan arc rather than a Jordan contour.) We map (λ) if it belongs to G, one-to-one conformally upon the interior of the upper half unit square $-1 < \mathfrak{x} < 1$, $0 < \mathfrak{y} < 1$ in the $(\mathfrak{z} = \mathfrak{x} + i\mathfrak{y})$-plane. One knows* that the mapping is one-to-one and continuous even including the boundary, and we can see to it that A, O, B map into $\mathfrak{z} = -1, 0, +1$, hence the arc AOB of Γ into the base of

*Carathéodory, Math. Ann. 73 (1913), 305-320; Conformal Representation, Cambridge Tract No. 28, Cambridge, Eng., 1932.

the rectangle.

All this is now applied to the vacuum H with its bounding Jordan contours Γ and γ. With every interior point of H we associate an admissible block. We may then cover the closure $H^{\cdot}$ of H by a finite number of such blocks and of boundary patches of the nature of (λ). We impose upon the admissible functions Ψ the conditions that they are of class D_1 in the interior of G, and that for each of the covering boundary patches Ψ as a function of the corresponding $\mathfrak{z}$ has continuous first derivatives in the representing rectangle including the open base $\mathfrak{y} = 0$, $-1 < \mathfrak{x} < 1$. The existence of admissible functions is secured by our above example because $\Psi = \sum \mu_q$ vanishes at all points sufficiently near to Γ and has the constant value 1 at all points sufficiently near to γ. We shall thus find a minimizing function which in each of the boundary patches can be continued as an odd harmonic function, $\phi(\overline{\mathfrak{z}}) = -\phi(\mathfrak{z})$, a little beyond the base into the fictitious lower half of the $\mathfrak{z}$-square. This will be so on Γ while continuation beyond the Jordan contours γ takes place according to the symmetry condition $\phi(\overline{\mathfrak{z}}) = 2 - \phi(\mathfrak{z})$.

On Γ we have $\frac{\partial \phi}{\partial \mathfrak{x}} = 0$, $\frac{\partial \phi}{\partial \mathfrak{y}} > 0$. Thus if we construct the analytic function $F = \phi - i\Theta$ in the unit square with the real part ϕ its derivative will not vanish along the real axis. (Incidentally F itself is a straightening parameter of the nature of $\mathfrak{z}$.) This enables us to ascribe a definite negative charge

$$\frac{1}{2\pi}\int d\Theta = -\frac{1}{2\pi}\int \frac{\partial \phi}{\partial \mathfrak{y}}\, d\mathfrak{x}$$

to any sufficiently small arc, and thus to any arc, of Γ. Moreover the differential dF, which is regular analytic in H, does not vanish near the boundaries and hence <u>the number of its zeros is finite</u>.

<u>We stick to the hypothesis that the fixed nucleus K_0 is bounded by analytic curves separating K_0 from $\overline{K}_0$; it</u>

is the boundary Γ of G which we permit to consist of Jordan contours of non-analytic character. We may then normalize $\phi = \frac{1}{c}\Phi$, $\frac{1}{c} = R$, in such a way that the charge $\frac{1}{2\pi}\int_\gamma d\vartheta$ on γ becomes unity.

Reversing the order we now put the formulas involving the voltage parameter r first. Since Γ_r is analytic for $0 \leqq r < R$ and $d\vartheta$ analytic on γ and Γ_r we obtain

$$\int_{\Gamma_r} d\vartheta = \int_\gamma d\vartheta = 2\pi,$$

$$N(r;\alpha) + m(r;\alpha) - m^0(\alpha) = T(r), \tag{4.2}$$

$$T(r) = \frac{1}{2\pi}\int \phi(r;\mathfrak{p})S(\mathfrak{p}). \tag{4.3}$$

With r tending to R, $N(r;\alpha)$ and $T(r)$ tend to

$$N(\alpha) = \sum \nu(\mathfrak{p};\alpha)\phi(\mathfrak{p}), \qquad T = \frac{1}{2\pi}\int \phi(\mathfrak{p})S(\mathfrak{p});$$

hence the equation (4.2) itself shows that the limit

$$\lim_{r\to R} m(r;\alpha) = \lim_{r\to R} \frac{1}{2\pi}\int_{\Gamma_r} \log \frac{1}{\|\alpha x\|}\, d\vartheta \tag{4.4}$$

exists, and if we denote it by $m(\alpha)$ the old equation (3.1) is reestablished. Similarly for all ranks p. The above remark about the charge on Γ shows that the limit of $\int_{\Gamma_r} d\vartheta$ for $r \longrightarrow R$ may be written as the Stieltjes integral $\int_\Gamma d\vartheta$, and the same is true for $m(\alpha)$,

$$m(\alpha) = \frac{1}{2\pi}\int_\Gamma \log \frac{1}{\|\alpha x\|}\cdot d\vartheta, \tag{4.5}$$

at least if no points of intersection between curve $\mathfrak{C}$ and plane (α) lie on Γ. But under all circumstances (4.5) can be interpreted as the limit (4.4).

Since there are no critical points near Γ, the line Γ_r is analytic in the strictest sense as soon as r is sufficiently near to R. Thus our result itself shows that $\mathfrak{R}$ can be exhausted by regions G with strictly analytic boundaries. On the other hand, we should like the concept of admissible G to be so wide as to include the G_r along with every G. For this reason we ultimately fix our assumptions as follows:

The nucleus K_0 is compact, bounded by a finite number of disjoint analytic contours γ which separate K_0 from $\overline{K}_0$; the complement $\overline{K}_0$ has no limited component.

Any open limited set $G \supset K_0$ is admissible if (1) there exists a continuous function Φ on $\mathfrak{R}$ which vanishes in $\overline{G}$, equals 1 in K_0, and is harmonic in $G - K_0$, and if (2) grad Φ vanishes in $G - K_0$ at a finite number of points only (critical points).

Our above considerations then show that these conditions are satisfied for a G bounded by a finite number of Jordan contours which have common points neither among themselves nor with γ, provided G has no component free from points of K_0.

Footnote explaining why preference has been given to the above procedure over against other methods suggested by the literature. (1) Adaptation of R. Nevanlinna's construction of harmonic measure, [10], pp. 24-25, would consist in three steps: construction of the universal covering surface $\tilde{H}$ of H, its conformal representation by the circle $|z| <$, application of Poisson's formula. The Dirichlet principle is one of the best, if not the best method for the middle part: construction of the uniformizing variable z. A further and somewhat devious argument is needed to prove that the conformal image of $\tilde{H}$ is the z-circle and not the plane. We have avoided not only the first topological part, which is clearly alien to the electrostatic problem for H, but also settled by a local rather than a global argument this intrinsically local question whether the arcs of the Jordan contours are mapped into segments or contracted into points. (2) R. Courant succeeds in proving directly by Dirichlet's principle that the minimizing function actually assumes the wanted boundary values along Jordan contours, without invoking Carathéodory's theorem (R. Courant and D. Hilbert, Methoden der mathematischen Physik, vol. 2, 1937, pp. 495-497). However it will probably not be easy to show in the same manner that the critical points do not cluster toward the boundary. Anyhow, this hard work was done before in the theory of conformal representation of plane domains, where it rightly belongs.

§5. Second main theorem

We propose to pave the way to the second main theorem by a function-theoretic definition of the Euler characteristic.

Let G be a compact part of the Riemann surface bounded by a finite number of disjoint analytic Jordan contours Γ, and dW a meromorphic differential in G, no zeros or poles of which lie on Γ. For a certain neighborhood of a boundary point $\mathfrak{p}_0$ we may use $\int_{\mathfrak{p}_0}^{\mathfrak{p}} dW = W(\mathfrak{p})$ as local parameter. The tangent of the image of the boundary contour in the W-plane will turn by a certain angle $\Delta(\arg dW)$ while the point $\mathfrak{p}$ describes an arc $\Delta = (\mathfrak{p}_1 \mathfrak{p}_2)$ of the contour contained in that neighborhood. This enables us to form the total increment

$$\frac{1}{2\pi}\int_{\Gamma} d\,(\arg dW)$$

along Γ described in the positive sense around G. This total increment clearly is an integer. Let $\nu(\mathfrak{p};dW)$ be the order of dW at any point $\mathfrak{p} \in G$ and

$$n_G(dW) = \sum \nu(\mathfrak{p};dW)$$

the total order of the differential in G.

LEMMA 5. A. *The difference*

$$\chi = \chi[G] = n_G(dW) - \frac{1}{2\pi}\int_{\Gamma} d(\arg dW) \tag{5.1}$$

is independent of the differential dW. This integer is called the Euler characteristic of G.

Indeed, let dW' be another meromorphic differential in G which has neither zeros nor poles on Γ. Apply (1.10) to the harmonic function $\log\left|\frac{dW'}{dW}\right|$ after cutting out its singular points by little contours. The result is the equation

$$\frac{1}{2\pi}\int_{\Gamma} d\left(\Im \log \frac{dW'}{dW}\right) = n_G(dW'/dW)$$

which proves the independence of (5.1) from dW.

We return to any admissible open G and the condenser $H = G - K_0$. It is clear that, for a non-critical value of r, the Euler characteristic of $H_r = G_r - K_0$ is the difference of the Euler characteristics $\chi(r)$ of G_r and χ_0 of K_0. Application of (5.1) to H_r and the differential $d\zeta = dW$ yields at once the fact that the number of critical points ϖ in $G_r - K_0$ equals $\chi(r) - \chi_0$. Indeed along the boundaries γ and Γ_r the differential $d\zeta$ is of constant argument $\pm \frac{\pi}{2}$. Because there is only a finite number of critical points in H, the characteristic $\chi(r)$ will have a constant value χ as soon as r is sufficiently near to R. Thus we can ascribe a definite Euler characteristic $\chi = \chi[G]$ to G itself, the difference $\chi[G] - \chi_0$ being the total number of critical points in H.

Our next move consists in applying Green's formula to the vacuum H, the potential ϕ, and the harmonic function

$$\phi_1 = \log \left|\frac{dZ}{d\zeta}\right|$$

where dZ is the familiar differential defined by

$$\frac{dZ}{dz} = \frac{(A^{p-1}X_z^{p-1}) \cdot (A^{p+1}X_z^{p+1})}{(A^p X_z^p)^2} .$$

The A's are special contravariant polyads. For simplicity's sake let us assume that dZ has no zeros or poles on γ. Properly taking the zeros and poles of $\frac{dZ}{dz}$ into account, we get the following relation

$$\text{(5.2)} \qquad \begin{aligned} &\int_{\Gamma_r} \log\left|\frac{dZ}{dz}\right| \cdot d\vartheta - \int_{\gamma} \log\left|\frac{dZ}{d\zeta}\right| \cdot d\vartheta \\ &\quad + r\int_{\gamma} d(\Im \log \frac{dZ}{d\zeta}) + 2\pi \sum_{\mathfrak{p}} \phi(r;\mathfrak{p})\nu^*(\mathfrak{p}) = 0 \end{aligned}$$

with the sum extending over all points $\mathfrak{p}$ in H_r and $\nu^*(\mathfrak{p})$ being the vanishing order $\nu(\mathfrak{p};dZ) - (e(\mathfrak{p})-1)$ of $\frac{dZ}{d\zeta}$ at $\mathfrak{p}$,

$\nu(\mathfrak{p};dZ) = \{\nu_{p-1}(\mathfrak{p};A^{p-1}) - 2\nu_p(\mathfrak{p};A^p) + \nu_{p+1}(\mathfrak{p};A^{p+1}\} +$

$\{v_p(\mathfrak{p})-1\}$. dZ is also defined in K_0 but $d\zeta$ is not. However because of the constant argument of $d\zeta$ along γ we have

$$\int_\gamma d(\Im \log \frac{dZ}{d\zeta}) = \int_\gamma d(\Im \log dZ),$$

and by our lemma this integral equals

$$2\pi m_{K_0}(dZ) - 2\pi\chi_0.$$

Addition of $r \cdot n_{K_0}(dZ)$ to the sum

$$\sum \phi(r;\mathfrak{p})\nu(\mathfrak{p};dZ)$$

extending over $\mathfrak{p} \in H_r$ results in the same sum extending over G_r or, what is the same, over the entire Riemann surface. Hence we are led to introduce the valence $V_p(r)$ of all stationary points of rank p in G_r,

$$V_p(r) = \sum \phi(r;\mathfrak{p}) \; \{v_p(\mathfrak{p})-1\}$$

and the quantity

$$\eta(r) = \chi_0 r + \sum \phi(r;\mathfrak{p})\{e(\mathfrak{p})-1\} \qquad (\mathfrak{p} \in H_r).$$

In the definition of $V_p(r)$ the summation over $\mathfrak{p}$ is unrestricted, whereas in $\eta(r)$ the restriction $\mathfrak{p} \in H_r$, or if one prefers $\mathfrak{p} \in H$, has to be imposed. It is also legitimate to write

$$\eta(r) = \chi_0 r + \sum \phi(r;\varpi),$$

if, as usually, ϖ runs over the critical points in H each counted with its proper multiplicity. With these notations (5.2) changes into the equation

$$\begin{aligned} &\{N_{p-1}(r;A^{p-1}) - 2N_p(r;A^p) + N_{p+1}(r;A^{p+1})\} + V_p(r) = \\ (5.3)\quad &\frac{1}{2\pi}\{\int_{\Gamma_r} \log|\frac{dZ}{d\zeta}| \cdot d\vartheta - \int_\gamma \log|\frac{dZ}{d\zeta}| \cdot d\vartheta\} + \eta(r). \end{aligned}$$

It is not difficult to remove the hypothesis that dZ is free from zeros and poles on γ and to extend the formula (5.3) to the critical values of r. In this connection observe that the definition of $\eta(r)$ is indifferent to whether or not one includes in the sum the critical points on Γ_r, because $\phi(r;\mathfrak{p})$ vanishes there. The function $V_p(r)$ is of regular type.

Now add

$$m_{p-1}(r;A^{p-1}) - 2m_p(r;A^p) + m_{p+1}(r;A^{p+1})$$

to both sides of the equation (5.3) and introduce the quantities

$$(5.4)\quad 2\Omega_p(r) = \frac{1}{2\pi}\int_{\Gamma_r} \log S^p_\vartheta \cdot d\vartheta,\quad 2\Omega^0_p = \frac{1}{2\pi}\int_\gamma \log S^p_\vartheta \cdot d\vartheta.$$

A formula results from which the auxiliary A's have disappeared:

$$(5.5_r)\quad \{T_{p-1}(r)-2T_p(r)+T_{p+1}(r)\} + V_p(r) = \{\Omega_p(r)-\Omega^0_p\} + \eta(r).$$

Finally let r tend to R. Then the three T(r) in (5.5_r) tend to T, $V_p(r)$ to the valence of the stationary points in G,

$$V_p = \sum_{\mathfrak{p}}\phi(\mathfrak{p})\ (v_p(\mathfrak{p}) - 1),$$

and $\eta(r)$ to

$$\eta = \eta[G] = \chi_0 R + \sum\phi(\mathfrak{w}),$$

the latter sum extending over all critical points $\mathfrak{w}$ in H. (It is here that the finiteness of their number becomes important.) Therefore $\Omega_p(r)$ must tend to a limit $\Omega_p = \Omega_p[G]$ and we get the

FORMULA OF THE SECOND MAIN THEOREM:

$$(5.5)\quad \{T_{p-1} - 2T_p + T_{p+1}\} + V_p = (\Omega_p - \Omega^0_p) + \eta.$$

Two expressions for η are available

$$(5.6) \qquad \eta = \chi_0 R + \sum \phi(\varpi) = \chi \cdot R - \sum (R - \phi(\varpi)).$$

They show that

$$(5.7) \qquad \chi_0 \leqq \eta / R \leqq \chi.$$

Application of (5.5) to G_r instead of G carries us back to the equation (5.5_r).

Ω_p is the familiar compensating term for the stationary points of rank p. A new feature is the appearance of the term η which depends on the configuration K_0, G but neither on the curve $\mathfrak{C}$ nor on the rank p. In the special case studied in Chaps. II and III, K_0 is a fixed circle in the z-plane and there are no critical points; hence $\chi_0 = -1$, $\eta = -R$ in complete agreement with III, §6. Comparison with the corresponding formula for algebraic curves exhibits the analogy between the term η here and the term 2g - 2 (g = genus) there. One knows that 2g - 2 has a purely topological significance. So has the Euler characteristic χ.

In order to show this we must make use of dissections rather than coverings, simply because topology in its present state is still dependent on that device. Hence in trying to evaluate the integral at the right of equation (5.1) let us suppose G to be triangulated, the dividing lines and the entire contour Γ of G having continuous tangents and avoiding the zeros and poles of dW. Analyticity of Γ is no longer required. But we assume that each triangle lies inside a conformal circle $|z| < a$. The integral of d(arg dW) along Γ equals the sum of the integrals of the same integrand extending over the contours Δ of the individual triangles; in its turn each of these integrals is the sum of the integrals along the three sides of the triangle. But

$$\int_\Delta d(\arg dW) = \int_\Delta d(\Im \log \frac{dW}{dz}) + \int_\Delta d(\arg dz).$$

According to Cauchy's integral theorem, the first term equals 2π times the total order of $\frac{dW}{dz}$ inside the triangle Δ. (This would be true even if the line over which we integrate did not consist of pieces with continuous tangents; simple continuity would be enough. Not that the path bounds something, but that it lies inside the simply connected circle $|z| < a$ is what matters here; cf. H. Weyl, Die Idee der Riemannschen Fläche, p. 67.) Let α, β, γ denote the angles of the triangle Δ in the z-plane. Then

$$\int_{\Delta} d(\arg dz) + (\pi-\alpha) + (\pi-\beta) + (\pi-\gamma) = 2\pi,$$

$$\int d(\arg dz) = (\alpha+\beta+\gamma) - \pi. \tag{5.8}$$

Let h_0, h_1, h_2 be the numbers of inner vertices, sides and triangles of the combinatorial pattern, and $h_0' = h_1'$ the numbers of vertices and sides on the boundary Γ. After assignment of a local parameter z^* to one of the vertices, $\mathfrak{p}_0$, the angle α of one of the triangles at the vertex $\mathfrak{p}_0$ is the same whether measured in the z- or the z^*-plane. The sum of all angles around $\mathfrak{p}_0$ is 2π or π, according to whether $\mathfrak{p}_0$ is an inner or a boundary vertex. Hence by adding over all triangles equation (5.8) leads to

$$2\pi\chi = -\sum \int_{\Delta} d(\arg dz) = \pi h_2 - 2\pi h_0 - \pi h_0'. \tag{5.9}$$

But since every inner side belongs to two triangles, every boundary side to one, $h_0' + 2h_1 = 3h_2$ or $h_0' = 3h_2 - 2h_1$. Thus (5.9) finally yields the well-known <u>topological definition of the Euler characteristic</u>,

$$\chi = h_1 - h_0 - h_2.$$

§6. Positive and zero capacity. The "little" terms m^0 and Ω^0

Once in possession of the formulas for the first and second main theorems we can turn to their evaluation.

We shall find that our estimates depend on a fundamental distinction between two sorts of non-compact Riemann surfaces, those of positive and of zero capacity, or, as F. Klein would have said, of hyperbolic and parabolic types (leaving the term elliptic to describe the compact surfaces). In case the voltages $R = R[G]$ for all admissible G are bounded, let I be the least upper bound; in the opposite case set $I = \infty$. For any positive ϵ we can ascertain an admissible G^{ϵ} such that $R[G^{\epsilon}] \geq I - \epsilon$ or $1/\epsilon$ respectively. Then $R[G] > I - \epsilon$ or $> 1/\epsilon$ for every G which encloses a certain compact set, namely the closure of G^{ϵ}. We express this fact by saying that $R[G]$ tends to the limit I under exhaustion of the Riemann surface $\mathfrak{R}$ by G,

$$R[G] \longrightarrow I \qquad \text{for} \quad G \longrightarrow \mathfrak{R},$$

and speak of the two cases as those of positive capacity $1/I$ (I finite) and zero capacity ($I = \infty$).

A simply connected non-compact Riemann surface $\mathfrak{R}$ is conformally equivalent either to the whole z-plane or to its unit circle K_1; we maintain that this is so according to whether $\mathfrak{R}$ is of zero or of positive capacity. Indeed, the unit circle K_1 has positive capacity, whatever compact part of it the nucleus K_0 occupies, because the electrostatic problem has a solution for the condenser $K_1 - K_0$ in the z-plane. In order to show that the complete plane is of zero capacity for any compact nucleus K_0, enclose K_0 by a circle K_0', $|z| < \rho_0$, and let G be any circle $|z| < \rho$ of radius $\rho > \rho_0$. The voltage of $G - K_0$ is larger than the voltage $\log \frac{\rho}{\rho_0}$ of $G - K_0'$, and the latter tends to infinity with $\rho \longrightarrow \infty$. It is desirable, and should not be too hard, to prove that the distinction between positive and zero capacity is independent of the nucleus K_0, even for multiply connected surfaces. But we shall not follow up this question here; our standpoint throughout has been that the structure underlying our investigation is a Riemann surface $\mathfrak{R}$ plus a definite nucleus K_0 on it.

$\psi_0[G]$ being a positive function of G and $\psi[G;\alpha]$ a function of G which may involve some parameters α, let the limit equation

$$\psi = O_+(\psi_0) \qquad \text{(uniformly in } \alpha)$$

express that a number B and a compact set K can be ascertained such that

$$\psi[G;\alpha] \leqq B\cdot\psi_0[G]$$

for all $G \supset K$ and all values of the parameters α. Notice the one-sidedness of the inequality! For a non-negative ψ the symbol O_+ has the same meaning as O. Similarly

$$\psi = o_+(\psi_0) \qquad \text{(uniformly in } \alpha)$$

states that for every $\epsilon > 0$ there exists a compact set K^ϵ such that

$$\psi[G;\alpha] \leqq \epsilon\,\psi_0[G]$$

for all $G \supset K^\epsilon$ and all α.

Before we can proceed further we have to investigate the inconspicuous terms $m^0[G;\alpha]$ and $\Omega_p^0[G]$ in the formulas of the first and second main theorems. In case the given plane $(\boldsymbol{\alpha})$ does not intersect $\mathfrak{C}$ on γ, $\|\alpha x\|$ has a positive lower bound $b(\leqq 1)$ on γ and therefore

$$0 \leqq m^0[G;\alpha] = \frac{1}{2\pi}\int_\gamma \log\frac{1}{\|\alpha x\|}\cdot d\vartheta \leqq \log 1/b.$$

The upper bound thus obtained depends on α though not on G. The argument fails for planes (α) which intersect $\mathfrak{C}$ on γ, and uniformity with respect to α is out of the question. Because of these shortcomings we proceed in a different way.

Let us fix an admissible G^0, of voltage R_0, and limit

ourselves to regions G containing G^0. For the two distributions $d\vartheta^0$, $d\vartheta$ which the condensers G^0, G induce on γ we have the inequality (2.5),

$$(0<)\ \frac{d\vartheta}{d\vartheta^0} \leqq \frac{R}{R_0}. \tag{6.1}$$

This makes it possible to appraise $m^0[G;\alpha]$ in terms of

$$m^0[G^0;\alpha] = \frac{1}{2\pi}\int_\gamma \log\frac{1}{\|\alpha x\|}\cdot d\vartheta^0$$

The lemmas II, 4. A and D show two things: (1) $m^0[G^0;\alpha]$ has an upper bound B which is independent of α; (2) for every $\epsilon > 0$ there exists a number B_ϵ such that the integral

$$\frac{1}{2\pi}\int_{\gamma_\epsilon} \log\frac{1}{\|\alpha x\|}\cdot d\vartheta^0$$

extended over the part $\log\frac{1}{\|\alpha x\|} \geqq B_\epsilon$ of γ is less than ϵ, irrespective of the position of the plane (α). In view of (6.1) the first fact leads to the estimate

$$m^0[G;\alpha] \leqq \frac{R}{R_0}\, m^0[G^0;\alpha] \leqq \frac{B}{R_0}\cdot R.$$

Consequently

$$m^0[G;\alpha] = \begin{cases} O(1) & (I \text{ finite}) \\ O(R) & (I = \infty) \end{cases} \tag{6.2}$$

(uniformly in α). Splitting

$$m^0[G;\alpha] = \frac{1}{2\pi}\int_{\gamma-\gamma_\epsilon} \log\frac{1}{\|\alpha x\|}\cdot d\vartheta + \frac{1}{2\pi}\int_{\gamma_\epsilon} \log\frac{1}{\|\alpha x\|}\cdot d\vartheta$$

and applying (6.1) to the second part, we obtain

$$m^0[G;\alpha] \leqq B_\epsilon + \frac{\epsilon}{R_0}\cdot R.$$

Hence in case of zero capacity, (6.2) may be improved to

$$m^0[G;\alpha] = o(R).$$

Nor is a better result required, since for zero capacity the relation $T_p \geqq c_p R$, derived from (3.10), shows that T[G] is at least of the order of magnitude of R[G]. The same argument as in Chap. III, §6, yields the asymptotic relation

$$R - T_p = o_+(R) \tag{6.4}$$

besides the inequality $T_p \geqq c_p R$. In the case of positive capacity, however, the first main theorem remains a significant statement only if $T_p[G] \to \infty$ as G exhausts $\mathfrak{R}$. Let us agree, therefore, that <u>all statements about curves in this case are made under the tacit assumption that</u>

$$[\infty] \qquad T_p[G] \longrightarrow \infty \text{ with } G \longrightarrow \mathfrak{R}.$$

Later on we need the corresponding properties for the quantities

$$\tilde{m}_p^0[G;A^h] = \frac{1}{2\pi}\int_\gamma \log \frac{1}{\|A^h : X^p\|}\cdot d\vartheta \qquad (h+p \leqq n)$$

namely

$$\tilde{m}_p^0[G;A^h] = O(1) \text{ (pos. cap.)}, = o(R) \text{ (zero cap.)} \tag{6.5}$$

uniformly with respect to all h-elements $\{A^h\}$. Let $a_1, \ldots, a_{n-p}$ be any n - p covariant vectors satisfying the conditions of a normal basis

$$(a_i | a_k) = \delta_{ik} \quad (\text{for } i,k = 1,\ldots,n-p). \tag{6.6}$$

Set

$$A^{n-p} = [a_1 \ldots a_{n-p}],\ A^{n-p-1} = [a_2 \ldots a_{n-p}], \ldots, A^1 = a_{n-p}.$$

Then

$$|X_\vartheta^p| \geqq |[A^1 X_\vartheta^p]| \geqq \ldots \geqq |[A^{n-p} X_\vartheta^p]|.$$

Hence it suffices to prove our proposition for $h = n - p$. In this case $[A^{n-p}X^p_\vartheta]$ has only one ordered component, namely the determinant

(6.7) $$|a_1, \ldots, a_{n-p}, x, \frac{dx}{d\vartheta}, \ldots|$$

which is an analytic function on γ, and thus the function-theoretic lemmas of II, §4 again become applicable. The argument is not that the components of X^p are linearly independent (which might not be so), but that for no (n-p)-uple of vectors $a_1, \ldots, a_{n-p}$ satisfying the conditions (6.6), the analytic function (6.7) vanishes identically. These (n-p)-uples form a compact manifold.

The treatment of Ω^0_p is complicated by the circumstance that now even the integrand $\log S^p_\vartheta$ along γ depends on the distribution $d\vartheta$ and hence on G. Fortunately this dependence is controlled by the simple law

$$S^p_\vartheta\left(\frac{d\vartheta}{d\vartheta^0}\right)^2 = S^p_{\vartheta^0}.$$

Dropping the index p we have

$$\log(1/S_\vartheta) \leqq \log(1/S_{\vartheta^0}) + 2 \log R/R_0,$$

consequently

$$-2\Omega^0[G] = \frac{1}{2\pi}\int_\gamma \log(1/S_\vartheta)\cdot d\vartheta \leqq \frac{1}{2\pi}\int \log(1/S_{\vartheta^0})\cdot d\vartheta + 2 \log \frac{R}{R_0}.$$

The first part on the right side may now be treated as $m^0[G;\alpha]$ before, though absence of any parameters α simplifies matters considerably. One surrounds the zeros of S_{ϑ^0} on γ by small arcs γ_ϵ such that $S_{\vartheta^0} \leqq 1$ on γ_ϵ and

$$\frac{1}{2\pi}\int_{\gamma_\epsilon} \log(1/S_{\vartheta^0})\cdot d\vartheta^0 < \epsilon.$$

Outside γ_ϵ, the integrand $\log(1/S_{\vartheta^0})$ will have an upper bound B_ϵ. Therefore

$$\frac{1}{2\pi}\int_\gamma \log(1/S_{\vartheta^0})\cdot d\vartheta \leqq B_\epsilon + \frac{\epsilon}{R_0} R.$$

The result is

$$(6.8)\qquad -\Omega^0[G] = \begin{cases} O_+(1) & \text{for pos. cap.} \\ o_+(R) & \text{" zero "} \end{cases}$$

§7. The fundamental inequality

After these preliminaries we establish the fundamental inequality* for $\Omega_p[G]$ in analogy to Chap. III, §6. Comparing the definition of Ω,

$$2\Omega_p(r) = \frac{1}{2\pi}\int_{\Gamma_r} \log S_\vartheta^p \cdot d\vartheta$$

with the expression for $Q_p(r)$ and observing that $d\vartheta > 0$ along Γ_r and $\int_{\Gamma_r} d\vartheta = 2\pi$, we find

$$2\Omega_p(r) \leqq \log Q_p(r) \text{ or } e^{2\Omega_p(r)} \leqq Q_p(r),$$

and thus (3.10) gives

$$c_pR + \int_0^R (R-r)\cdot e^{2\Omega_p(r)} dr \leqq T_p.$$

Anticipating later applications, we prefer from the outset the second of the viewpoints discussed in III, §6. For zero capacity

$$1 + (1 - c_p)R \leqq (1 - \tfrac{1}{2}c_p)R \leqq T_p.$$

as soon as $R \geqq 2/c_p$ and $(1 - \frac{1}{2}c_p)R \leqq T_p$ [see Eq. (6.4)]; therefore

*Cf. H. and J. Weyl, Proc. Nat. Ac. 28, 1942, 417-421. The essential idea which makes the estimates click is the introduction of $d\zeta$ as the differential against which all other differentials are measured. On p. 420 l. c. the limit potential ϕ should not have been used; in either case, that of zero or of positive capacity, one must operate with the potential ϕ_G of the exhausting region G. Whether in the second case the limiting ϕ works is a difficult question, on which Fatou's theorem throws some light, but which need not concern us here.

$$(7.1) \qquad 1 + R + \int_0^R (R-r)e^{2\Omega_p(r)}dr \leqq 2T_p$$

for all G containing a certain compact K. In case of positive capacity, the assumption [∞] allows us to write

$$1 + (1-c_p)R \leqq 1 + I = o(T_p),$$

and the relation (7.1) holds even if the factor 2 in the right member is replaced by any number > 1.

We find it convenient to introduce a symbol $\omega = \omega_B$ involving the "parameter" B. For any two functions s[G], T[G] of G the equation $s = \omega_B(T)$ means that there is a compact set K such that

$$(7.2) \qquad 1 + R + \int_0^R (R-r)e^{s(r)}dr \leqq BT$$

for all admissible $G \supset K$; $T = T[G]$, $R = R[G]$, $s(r) = s[G_r]$. Continuity of $s(r)$ for $0 \leqq r < R$ is assumed. Clearly $s = \omega_B(T)$ implies $s_1 = \omega_B(T)$ for any function $s_1 \leqq s$ and $s_2 = \omega_{B'}(T)$ for $s_2 = s + c$, c being any positive constant and $B' = B \cdot e^c$. We may now state the relation (7.1) in the abbreviated form

$$(7.3) \qquad 2\Omega_p = \omega_2(T_p).$$

In evaluating the relation $s = \omega(T)$ in both cases, that of zero and positive capacity, we generalize our former procedure by introducing a continuous positive function $\lambda(r)$ in the interval $0 \leqq r < I$ <u>such that</u> $\int_0^I \lambda(r)dr$ <u>diverges</u>. The integral

$$\int_{r_1}^{r_2} \lambda(r)dr = \mu(r_1 r_2) \qquad (0 \leqq r_1 < r_2 < I)$$

serves as the (λ-) measure of any subinterval (r_1, r_2) of $(0, I)$. Our former choice $\lambda(r) = 1$ is good for zero, but not for positive capacity. We propose to take $\lambda(r) = r^h$ with some exponent $h \geqq 0$ for infinite I, and $\lambda(r) = (I-r)^{-h}$ with some exponent $h \geqq 1$ for finite I. Instead of Lemma III, 6. A. we now obtain

LEMMA 7. A. The relation $s = \omega_B(T)$ implies that the inequality

(7.4) $$s(r) \leqq \kappa^2 \log T(r) + (1+\kappa) \log \lambda(r) + d$$

holds for all admissible G and $0 < r < R$ under the condition that $G_r \supset K$ and r does not belong to a certain open subset of the interval $0 < r < R$ of measure less than

$$M = \frac{2}{\kappa - 1} \cdot e^{-d'} \qquad \{(1+\kappa)d' = d - \kappa^2 \log B\}.$$

Here $\kappa > 1$ and d may be chosen at random.

Indeed, under the assumption $f(0) \geqq 1$ the inequality

$$\frac{df}{dr} > e^{d'} \cdot \lambda(r)(f(r))^{\kappa}$$

will hold in an open set, the measure $\int \lambda(r)dr$ of which must be less than

$$e^{-d'} \int_1^{\infty} y^{-\kappa} dy = e^{-d'} \cdot \frac{1}{\kappa - 1}.$$

Set

$$1 + r + \int_0^r (r-\rho)e^{s(\rho)} d\rho = t(r)$$

and apply this remark first to $f(r) = \frac{dt}{dr}$ and then to $f(r) = t(r)$. Observing that the hypothesis (7.2) for G_r instead of G implies $t(r) \leqq B \cdot T(r)$ as soon as $G_r \supset K$, we find that under this hypothesis, and excepting an open set of measure less than $\frac{2}{\kappa - 1} e^{-d'}$,

$$s(r) \leqq \kappa^2 \log (B \cdot T(r)) + (1+\kappa)(\log \lambda(r) + d')$$

Given G, those r for which $G_r \supset K$ form a certain open interval $R_K < r < R$ ending at R (which of course will be empty unless $G \supset K$.) It is remarkable that the upper bound M for the measure of the exceptional set is independent of the choice of $\lambda(r)$, and in the application of the lemma to (7.3) also from the curve $\mathfrak{C}$.

Let us set

$$(7.5)\quad 2L(r) = (1+\kappa)\log \lambda(r) = \begin{cases} (1+\kappa)\cdot h \log r & (I = \infty), \\ (1+\kappa)\, h \log \frac{1}{I-r} & (I \text{ finite}). \end{cases}$$

The quantity that matters in the formula for the second main theorem is the difference $\Omega_p - \Omega_p^0$, and since $-\Omega_p^0[G] = o_+(R)$ is all we can guarantee in the case of zero capacity, we lose little by choosing a <u>high</u> exponent for h in $\lambda(r) = r^h$, but we gain inasmuch as the sets of fixed λ-measure M will be the smaller, the higher h is. In case of positive capacity, the term $L(r)$ is a necessary evil, and again a high h in $\lambda(r) = (I-r)^{-h}$ is on the whole advantageous.

After having fixed $\lambda(r)$ we introduce the phrase "<u>a function</u> $f[G]$ <u>satisfies the inequality</u> $f[G] \leq 0$ <u>for almost all</u> G" with the following meaning: <u>There exists a positive number</u> M <u>and a compact set</u> K <u>such that for every admissible</u> G <u>those</u> r <u>in the interval</u> $R_K < r < R$ (defined by $G_r \supset K$) <u>for which</u> $f[G_r] > 0$ <u>form an open subset of measure less than</u> M. Inequalities holding for almost all G will be marked by $||$. Replacing κ^2 by κ we have then proved:

LEMMA 7. B. $s = \omega(T)$ implies the fact that for any constants $\kappa > 1$ and d the inequality

$$(7.6)\qquad || \qquad s \leq \kappa \log T + L(R) + d$$

holds for almost all G.

Because the join of several open sets of measures less than $M_1, \ldots, M_m$ is one of measure less than $M = M_1 + \ldots + M_m$ we can state the <u>combination principle</u>:

LEMMA 7. C. Given a finite number of inequalities $f_i[G] \leqq 0$ the fact that each is satisfied for almost all G implies that the same is true for them simultaneously.

Having proved an inequality $f(r) \leqq 0$ for all r in the interval $R_K < r < R$ except in a set of measure less than M, we should like to conclude that the interval (R',R) of measure $\mu(R'R) = M$ contains values r for which $f(r) \leqq 0$. The interval (R',R) will exist as soon as $\mu(0R) > M$. The only doubtful point is whether $G_{R'}$ contains K and thus $R' > R_K$. We prove that this is so as soon as G is sufficiently large, i. e. as soon as G encloses a properly chosen compact part of $\mathfrak{R}$. Take an admissible $G^0 \supset K$. The function Φ_{G^0} will have a <u>positive</u> minimum δ in K,

$$\Phi_{G^0}(\mathfrak{p}) \geqq \delta \qquad \text{for } \mathfrak{p} \in K.$$

<u>A fortiori</u>

$$\Phi_G(\mathfrak{p}) \geqq \delta \text{ or } \phi_G(\mathfrak{p}) \geqq \delta R \text{ for } G \supset G^0 \text{ and } \mathfrak{p} \in K.$$

In other words, $G \supset G^0$ implies that the "trimmed" $G_{(1-\delta)R}$ contains K. But for our choice of $\lambda(r)$ it is not only true that the total measure of 0I is infinite, but that for any given $\sigma < 1$

$$\mu(\sigma R, R) = \int_{\sigma R}^{R} \lambda(r)dr \longrightarrow \infty \text{ with } R \longrightarrow I.$$

Thus $G_{R'}$ will contain K if $G \supset G^0$ and $\mu((1-\delta)R,R) > M$.

G_r arises from G by trimming, the width of the rim taken off being $\mu(rR)$. We may now state

LEMMA 7. D. Suppose the inequalities $f_i[G] \leqq 0$ $(i = 1,\ldots,m)$ satisfied for almost all G. Then there exists a number M such that each sufficiently large admissible G can be changed by cutting off a rim of width less than M into a region G^* $(=G_r, \mu(rR) < M)$

for which all the inequalities $f_1[G^*] \leqq 0$ are satisfied. If G varies so as to exhaust $\mathfrak{R}$ e. g. by running over a sequence $G^{(1)}, G^{(2)}, \ldots$, then the trimmed G^* exhausts $\mathfrak{R}$.

(The higher the exponent h of the measure function $\lambda(r)$ the less perceptible will be the trimming for large G.)

LEMMA 7. E. Suppose that for any given positive constant ϵ the inequality $f[G] < \epsilon$ holds for almost all G. Then

$$\underline{\lim}\, f[G] \leqq 0.$$

Proof. Evident.

This fact remains true even if we limit G to what we describe as a <u>complete system $\mathfrak{G}$ of admissible regions</u>. Completeness requires (1) that for any compact K there exist regions $G \in \mathfrak{G}$ containing K ($\mathfrak{R}$ is exhaustible by regions G of the system $\mathfrak{G}$), and (2) that G_r $(0 < r \leqq R)$ is in $\mathfrak{G}$ if G is. The phrase "<u>for almost all</u> G" is also used in the restricted sense "<u>for almost all</u> G <u>in a given complete set</u> $\mathfrak{G}$". One of the simplest implications of the relation $f[G] \leqq 0$ holding for almost all G is the fact that the opposite inequality $f[G] > 0$ cannot prevail for all sufficiently large G in a given complete set.

We are now ready to return to the formula of the second main theorem (5.5). Let us abbreviate by (o) any non-negative quantity depending on G which is O(1) for finite I and o(R) for infinite I. In view of the assumption [oo] for finite I and of the relation $R = O(T_p)$ for infinite I we have in both cases $(o) = o(T_p)$. The difference $\Omega_p - \Omega_p^0$ may be written as $\Omega_p' + o_p$ where

$$\Omega_p' = \Omega_p,\ o_p = -\Omega_p^0 \qquad \text{if } \Omega_p^0 \leqq 0, \text{ and}$$

$$\Omega_p' = \Omega_p - \Omega_p^0,\ o_p = 0 \quad \text{if } \Omega_p^0 \geqq 0.$$

o_p is a quantity of type (o) and $\Omega'_p \leqq \Omega_p$, thus $\Omega'_p = \frac{1}{2}\omega(T_p)$ and

$$(7.7) \qquad \Omega_p - \Omega_p^0 = (o) + \tfrac{1}{2}\omega(T_p).$$

Suppose we know that T_p grows faster than η, and also faster than $\log \frac{1}{I-R}$ in the case of finite I. Then (7.7) implies that, given any $\epsilon > 0$, the left member of (5.5) is less than $\epsilon \cdot T_p$ for almost all G. As a matter of fact it is sufficient to assume the condition just mentioned to be satisfied for almost all G. Hence we put down the following

HYPOTHESIS $\mathfrak{H}$: Given any positive ϵ, the inequalities

$$(\mathfrak{H}') \quad \| \qquad \eta < \epsilon \cdot T_p \qquad (p = 1,\ldots,n-1)$$

hold for almost all G, and in case I is finite, the same is true for the inequalities

$$(\mathfrak{H}'') \quad \| \qquad \log \frac{1}{I-R} < \epsilon \cdot T_p.$$

We speak of the <u>restricted</u> <u>hypothesis</u> $\mathfrak{H}$ if we limit ($\mathfrak{H}'$) and ($\mathfrak{H}''$) to regions of G of a given complete system $\mathfrak{G}$. Since T_p is an increasing function of G, ($\mathfrak{H}''$) even in the restricted sense implies [∞].

As in III, §7, we conclude:

THEOREM. Let $\epsilon > 0$ be given. Under the hypothesis $\mathfrak{H}$ the inequalities

$$\| \; T_{p+1} < (1 + \frac{1}{p} + \epsilon)T_p, \quad T_{p-1} < (1 + \frac{1}{n-p} + \epsilon)T_p,$$

$$\| \qquad V_p < (\frac{n}{p(n-p)} + \epsilon)T_p$$

$$\| \qquad V_p + (T_{p+1} + T_{p-1}) < (2 + \epsilon)T_p$$

will hold for almost all G.

Examples show that the hypothesis $\mathfrak{H}$ is actually essential (cf. Chap. V, §3). It is no wonder that the structure of the Riemann surface should reveal itself in some way in the behavior of its analytic curves; we learn here that it does so primarily by way of the one quantity $\eta = \eta[G]$. The part ($\mathfrak{H}'$) of hypothesis $\mathfrak{H}$ is satisfied *a fortiori* if for any $\epsilon > 0$ the Euler characteristic $\chi = \chi[G]$ fulfills the inequality

$$\| \qquad \chi < \epsilon \cdot \frac{T_p}{R}$$

for almost all G.

Above we have pleaded for a high value of the exponent h in (7.5). However the lowest possible value, $h = 0$ and $h = 1$ respectively, is indicated if we care for nothing but the fact that the resulting inequalities hold for almost all G, disregarding how big or small the exceptional r-sets turn out. Pure theorists that we are, we propose to abide by this convention hereafter, and thus write

$$(7.8) \quad L_\kappa = 0 \ (I = \infty), \quad L_\kappa = \frac{1+\kappa}{2} \log \frac{1}{I-R} \ (I \text{ finite}).$$

The assumption that really mattered in the proof of our last theorem is the inequality

$$(7.9)_p \quad \| \qquad \eta + L_\kappa < \epsilon \cdot T_p$$

holding for almost all G after $\epsilon > 0$ is arbitrarily fixed. In case of positive capacity it is obvious from the combination principle that $(7.9)_p$ is implied in the two simultaneous inequalities ($\mathfrak{H}'$), ($\mathfrak{H}''$), but the converse is less obvious because we are not sure that η is positive. However, from $\eta \geqq \chi_0 R$ there follows that $\eta \geqq \eta_0$ where the constant $\eta_0 = 0$ if $\chi_0 \geqq 0$ and $\eta_0 = \chi_0 I$ if $\chi_0 < 0$. Hence

$$(7.10) \qquad \log \frac{1}{I-R} \geqq 0, \qquad \eta \geqq -\frac{1}{2} \log \frac{1}{I-R}$$

as soon as R is sufficiently near its upper bound I, and then $(7.9)_p$ implies

$$\| \qquad \eta < \epsilon T_p, \qquad \frac{\kappa}{2} \log \frac{1}{I-R} < \epsilon T_p.$$

It is therefore perfectly legitimate in either case of zero and of positive capacity to formulate the hypothesis $\mathfrak{H} = \mathfrak{H}_p$ for the rank p by $(7.9)_p$.

A last remark is to the effect that, granted $[\infty]$, the hypothesis $\mathfrak{H}$ <u>for one</u> p is sufficient; it will then be automatically satisfied for all p. Indeed, without any further assumption than $[\infty]$, we have

$$(7.11) \quad \| \quad V_p + (T_{p+1} + T_{p-1}) < (2 + \epsilon)T_p + \eta + L_\kappa.$$

Hence $(7.9)_{p-1}$ implies

$$\| \qquad (1 - \epsilon)T_{p-1} < (2 + \epsilon)T_p$$

and consequently

$$\| \qquad \eta + L_\kappa < \frac{\epsilon(2+\epsilon)}{(1-\epsilon)} T_p,$$

which amounts to $(7.9)_p$. Applying the combination principle in the same manner to (7.11) and $(7.9)_{p+1}$, we see that the hypothesis $\mathfrak{H}$ carries over not only from p-1 to p but also from p+1 to p.

§8. Special cases of rotational symmetry

A characteristic feature of the Riemann surface $\mathfrak{R}_0$ underlying the investigations of Chaps. II and III is its <u>rotational symmetry</u>. Indeed, the z-plane may be described as a circle of infinite radius, and nothing but the desire to maintain this symmetry was responsible for our limitation to <u>circular</u> condensers. Other cases of rotational symmetry are the circle of finite radius and the annulus bounded by two concentric circles where the values ∞ and 0 for the radii of the outer and the inner circle respectively, are not excluded. In each of these instances the circle or annulus may be replaced by a Riemann surface which covers it with the same finite number of sheets everywhere possessing only isolated ramification points.* One chooses the nucleus K_0 so that the rotational symmetry is not destroyed, namely as a (small) circle or as a (thin) concentric annulus. And one limits oneself to symmetric regions G, namely to concentric circles and concentric annuli respectively. In the case of Riemann surfaces with several sheets, one replaces these figures by what covers them on the Riemann surface. The symmetric G clearly form a complete system $\mathfrak{G}$. Under these special circumstances the distribution of electricity over the circle γ will be uniform, and therefore not depend on the growing region G, and the terms $m^0(\alpha)$ and Ω_p^0, which caused some trouble in the general theory, are independent of G. In each of these cases it is easy to write down the potential ϕ and thus establish the theory in the same explicit form as attained in Chaps. II and III for the z-plane. In the circular case we have the formula (II, 1.7), whether the Riemann surface is a circle of infinite or of finite radius. In the annular case let the condenser $G - K_0$ be described by

$$G:\ R' < |z| < R;\quad K_0:\ r_0' \leqq |z| \leqq r_0 \quad (0<R'<r_0'<r_0<R).$$

Then

$$(8.1)\qquad \phi(z) = \begin{cases} e' \log \dfrac{|z|}{R'} & \text{for } R' \leqq |z| \leqq r_0', \\[2ex] e \ \log \dfrac{R}{|z|} & \text{for } r_0 \leqq |z| \leqq R \end{cases}$$

where the charges e', e are determined by

$$e' \log \frac{r_0'}{R'} = e \log \frac{R}{r_0}, \qquad e + e' = 1.$$

*E. Ullrich, Jour. f.d.reine u. angew. Math. 167, p. 198.

The circle has zero capacity if its radius is infinite, and so has the annulus if the outer radius is infinite and the inner radius 0; otherwise the capacity is positive. A factor $\frac{1}{h}$ is to be added to the right side of the formulas (II, 1.7) and (8.1) if circle or annulus is replaced by a Riemann surface covering it with the same finite number h of sheets everywhere; the (isolated) points of ramification are the critical points $\mathfrak{w}$ of the general theory.

In the older form of the theory of meromorphic functions a point charge was used instead of the conductor K_0. Even for the full z-plane the point charge had its disadvantages, but for an annulus this choice would completely destroy the rotational symmetry. It seems to have been for no other reason that whereas the theory of meromorphic functions in the z-plane had at once been generalized to the circle of finite radius, the annulus had to wait until quite recently. The example of the annulus also demonstrates that it would have been a mistake to restrict the nucleus K_0 by the assumption of simple connectivity.

§9. Appendix. The banks of a Jordan contour

THEOREM. A Jordan contour on $\mathfrak{R}$ has two banks.

As a first step we prove a lemma due to Carathéodory about Jordan contours Γ in a plane.* Let 0 be a point on Γ, r_0 a positive number such that there are points on Γ, e. g. 0', for which the distance $00' > r_0$. We determine an arc $\gamma = A0B$ of Γ which lies in the (closed) circle K_0 around 0 of radius r_0 and a positive number r_1 ($<r_0$) such that the distance $0z \geq r_1$ for every point z on the complementary arc $\gamma' = A0'B$.

*Mathem. Annalen 73, 1913, pp. 316-317.

LEMMA. If two points P, Q of distances OP, OQ $< r_1$ can be joined by a path not meeting Γ, then they may be joined by such a path inside K_0.

Proof. Both arcs OO' have a point in common with the periphery k_0 of K_0. Hence k_0 contains at least two distinct points of Γ. If the path joining P with Q does not lie entirely within K_0 it will meet its periphery k_0; suppose that this happens for the first time at the point P_1. The part P P_1 of the path lies in the circle K_0, and P_1 is situated on an open arc $\beta_1 = A_1B_1$ of k_0 of which

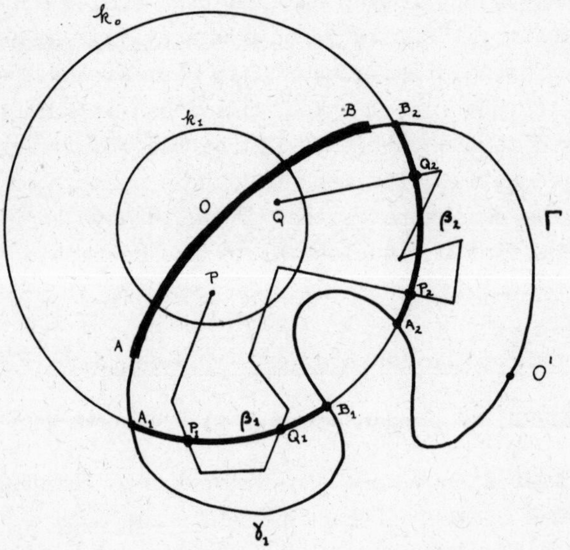

FIG. 3

the end points A_1, B_1 but no other points lie on Γ. Let Q_1 be the last point of the part P_1Q of the path which lies on β_1. We then replace the piece P_1Q_1 of the path by the sub-arc P_1Q_1 of β_1 (modified path). We propose to prove that immediately after Q_1 the path moves inside the circle K_0. Then it will remain in K_0 from then on, if it never meets k_0 again. In the opposite case there will be a first point P_2 after Q_1 where the path meets k_0, P_2 will be on another arc $\beta_2 = A_2B_2$ of k_0 of the same type as β_1 and the part Q_1P_2 of the path will lie in K_0. The end point P_2 is the first point of the modified path $PP_1Q_1P_2$ on β_2. Hence the same argument may be repeated and we finally obtain a modified path which consists of parts of the old path and arcs of the circle k_0 and which satisfies the desired conditions. It is easy afterwards to substitute broken lines for the circular arcs.

However, all this depends on the certainty that the path after Q_1 continues for a while inside and not outside K_0. Consider the Jordan contour Γ^1 formed by the circular arc β_1 and the arc $\gamma_1 = B_1A_1$ of Γ which is part of γ'. Each point z not on Γ^1 has a definite degree h with respect to Γ^1, namely the number of times Γ^1 winds around z in the positive sense. If the variable point z crosses the arc β_1 along a radius, its degree jumps by 1. As the circle K_1 of radius r_1 around 0 has no point in common with Γ^1, all its points z, in particular P and Q, have the same degree h. But if the path immediately after Q_1 moved outside the circle k_0, then the degree from there on would be $h + 1$ rather than h. To make this quite clear, use polar coordinates r, ϑ with the center 0 and determine an annular sector

$$\vartheta_0 < \vartheta < \vartheta_0', \qquad r_0 - \epsilon < r < r_0 + \epsilon$$

which contains the sub-arc P_1Q_1 of β_1 but no point of γ_1.

Then all points of the inner half $r_0 - \epsilon < r < r_0$ of the sector will have the same degree h, all points of the outer half $r_0 < r < r_0 + \epsilon$ the degree h + 1. We thus arrive at a contradiction.

After having established the lemma we continue our construction as follows. Choose a point L_0 on the arc OA of Γ and another point M_0 on the arc OB both distinct from O and at a distance from O less than r_1, and two points L', M' in the <u>exterior</u> G of Γ sufficiently close to L and M respectively. Then $OL' < r_1$, $OM' < r_1$, and according to our lemma L', M' may be joined by a path in G inside the circle K_0. After adding the segments $L'L_0$, $M'M_0$ but stopping on them when one first reaches Γ, one obtains a crosscut λ of G whose end points L, M on Γ are as close as one wishes to L_0 and M_0 and hence on the arcs OA and OB respectively. Γ itself is divided into the arc Γ_1 = LOM and Γ_2 = MO'L. The exterior G of Γ is divided by λ into two regions G_1, G_2 of which G_1 is bounded by the Jordan contour $\Gamma_1 + \lambda$ and G_2 by $\Gamma_2 + \lambda$. We denote by G_1, G_1' the two regions determined by $\Gamma_1 + \lambda$ while $\Gamma_2 + \lambda$ determines G_2, G_2'. Then the open arc Γ_2 lies in G_1' and the open arc Γ_1, in particular O, lies in G_2'. (These statements are integral parts of the proof of Jordan's theorem.*)

A circular neighborhood of O so small that it has no points in common with λ and Γ_2 will therefore lie entirely in G_2', and those of its points which belong to G are neither in G_2 nor on λ, hence in G_1. In other words, all points sufficiently near to O and not on Γ belong either to the interior G_0 of Γ or to the region G_1. It remains to be shown that G_1 is the interior (and not the exterior) of $\Gamma_1 + \lambda$. The arc AO'B contains a point O_1 of greatest distance from O, and this distance, like that of O', certainly exceeds r_0. O_1 lies in G_1' and as the prolongation

*See O. Veblen, Trans. Am. Math. Soc. 6, 1905, 93-98; L. E. J. Brouwer, Math. Annalen 69, 1910, 169-175.

of the radius OO_1 beyond O_1 cuts neither Γ nor λ, G_1' must be the exterior and G_1 the interior of $\Gamma_1 + \lambda$.

From the plane we turn to the Riemann surface $\mathfrak{R}$. Let Γ now be a Jordan contour on $\mathfrak{R}$. Let O be a point of Γ, z a local parameter for O, $|z| < a$, be a conformal neighborhood of O and $\gamma = A_0OB_0$ an arc of Γ which lies in this neighborhood. We may then choose a positive number a_1 ($<a$) such that the complementary arc $\gamma' = A_0O'B_0$ has no point in common with the circular disk $|z| \leqq a_1$. Choose points A', $B' \neq O$ inside the disk on the arcs $\widehat{OA}_0$, $\widehat{OB}_0$ of Γ and join them by a path (consisting of one or two segments) which is contained in the disk and avoids the

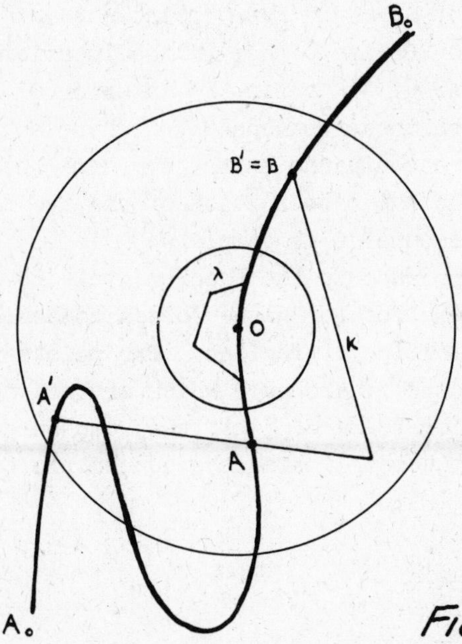

FIG. 4

point O. While z describes this path there will be a last point $z = A$ where it intersects with $\widehat{OA}_0$ and a first point B after A where it intersects $\widehat{OB}_0$. The arc $\Gamma_0 =$ AOB of Γ and the part $\kappa = AB$ of the path jointly constitute a Jordan contour Γ'. Finally choose a positive r_0 ($< a_1$) such that the arc AO'B does not penetrate into the disk $|z| \leqq r_0$. With the Jordan contour Γ' in the z-plane and this number r_0 carry out the construction described above. Since the crosscut $\lambda = LM$ to which it leads does not leave the circle of radius r_0 around O it cannot intersect the arc AO'B of Γ, and hence only its end points L, M are on Γ. As before, denote by Γ_1 the arc LOM of Γ (which is part of Γ_0). Every point not on Γ and sufficiently near to O lies either in the interior G_0 of the Jordan contour $\Gamma_0 + \kappa$ or in the interior G_1 of the contour $\Gamma_1 + \lambda$. Both these regions G_0, G_1 have no point in common and are contained in the conformal neighborhood $|z| < a$. Thus the construction of the two banks of Γ in the neighborhood of O is complete.

Another approach to the theorem could be attempted by repeating the proof for plane Jordan contours with this modified definition of regions: Two points P, Q in the circle of radius r_1 around O which are not on Γ belong to the same region (bank of Γ) if they can be joined without crossing Γ by a path inside the concentric circle of radius r_0.

CHAPTER V

THE DEFECT RELATIONS

§1. Picard's theorem

Picard's famous theorem states that a meromorphic function of z, $w = f(z)$, which is not a constant, assumes all values except at most 2. Should it leave out 3 distinct values, one could by means of a linear transformation of w locate these values at $w = 0, 1, \infty$. Picard's own proof made use of the inverse $\omega(t)$ of the modular function. The fundamental periods ω_1, ω_2 of an elliptic integral of the first kind $\int \frac{dz}{\sqrt{(P(z))}}$ in which $P(z)$ is a polynomial of formal degree 4 with 4 distinct roots of cross ratio t, can be so chosen that the imaginary part of $\omega = \omega_2/\omega_1$ is positive. Even so, ω is far from being uniquely determined because the choice of the fundamental periods in the whole lattice of periods $h_1\omega_1 + h_2\omega_2$ (h_1, h_2 arbitrary integers) involves a degree of arbitrariness which is expressed by the linear transformations of ω_1, ω_2 with integral coefficients of determinant 1 (modular group). But $\omega(t)$ is an analytic function in the Weierstrass sense which permits of analytic continuation along any path in the t-plane after the points $t = 0, 1, \infty$ have been removed. (These are the values which the cross ratio of 4 distinct points never assumes.) By a suitable linear transformation of ω, e. g. by using $\frac{\omega-i}{\omega+i}$ instead of ω, the upper half plane $\Im\omega > 0$ may be transformed into the interior of the unit circle $|\omega| < 1$.

The existence of this function may also be derived, and in several ways, from the possitility of mapping any simply connected region one-to-one and conformally onto the interior of the unit circle (or the upper half plane).

Now let f(z) be a meromorphic function which does not assume the values 0, 1, ∞ (i. e. an entire function which does not assume the values 0, 1). Weierstrass's monodromy theorem shows that $\omega(f(z))$ is a single-valued regular function in the whole z-plane. Because it satisfies the inequality $|\omega(f(z))| < 1$ it must be a constant and the same fate befalls f(z) itself.

Borel replaced Picard's proof by a more elementary yet considerably more complicated approach. Stated in terms of our quantities T and N his result amounted to something like this. Under the hypothesis that the integral $\int^{\infty} \frac{T(r)}{r^{\lambda+1}}\,dr$ diverges for a certain exponent λ there cannot be more than 2 values a for which $\int^{\infty} \frac{\tilde{N}(r;a)}{r^{\lambda+1}}\,dr$ converges. $\tilde{N}(R;a)$ is the valence of all points z in the circle of radius R where f(z) = a. We use our homogeneous form of writing, $f = x_1/x_2$, $a = a_1/a_2$ and, as before, denote by $\nu(z_0:a)$ the order in which $x_1a_2 - x_2a_1$ vanishes at z_0; then

$$\tilde{N}(R;a) = \sum_z \phi(R;z)\nu(z:a).$$

The exponent λ in Borel's proposition betrays its origin within the frame of the old theory of entire functions of finite genus. It would be more satisfactory if, by shedding these remnants, one could prove instead: There cannot be more than 2 distinct values a for which

$$\tilde{N}(R;a)/T(R) \to 0 \quad \text{as } R \to \infty. \tag{1.1}$$

R. Nevanlinna introduced the limit superior $1 - \delta(a)$ of the quotient (1.1) and called $\delta(a)$ the defect of the value a. Its vanishing indicates that the a-places of f do not permanently fall short of that density which the order T of f leads one to expect:

$$\overline{\lim}\ \tilde{N}(R;a)/T(R) = 1.$$

The extreme in the other direction is (1.1) or $\delta(a) = 1$. By methods which revolutionized this whole branch of the theory of analytic functions Rolf Nevanlinna proved that the sum of defects of any number of distinct values a may never exceed 2. This is clearly a very far-reaching generalization of Picard's theorem, and in particular implies the above conjecture. We shall follow here a somewhat different approach which was inaugurated by Rolf's brother Frithiof, and later greatly simplified by L. Ahlfors.

§2. Weighted averages. The basic general formula

Once more we consider a non-degenerate analytic curve $\mathfrak{C}$ stemming from the Riemann surface $\mathfrak{R}$ and write down the equation of the first main theorem for any admissible G:

$$N(\alpha) + m(\alpha) - m^0(\alpha) = T.$$

We introduce a positive function $\rho(\alpha) = \rho(\alpha_1, \ldots, \alpha_n)$ on the unit sphere, and form for any function $F(\alpha)$ on it the <u>weighted</u> <u>average</u>

$$\mathfrak{M}_\alpha' F = \int F(\alpha)\rho(\alpha)\cdot d\omega_\alpha / \int \rho(\alpha)\cdot d\omega_\alpha. \tag{2.1}$$

$\rho(\alpha)$ may have singularities, but the integral of $\rho(\alpha)$ over the unit sphere is supposed to be finite. In general it will not be true that the weighted averages $\mathfrak{M}'$ of $m(\alpha)$ and $m^0(\alpha)$ coincide as the ordinary averages do. But as $m(\alpha) \geqq 0$ implies $\mathfrak{M}' m(\alpha) \geqq 0$ we obtain the following inequality

$$\mathfrak{M}' N(\alpha) \leqq T + m_0', \tag{2.2}$$

which now takes the place of the equation IV, (3.3) and in which m_0' is the average

$$\mathfrak{M}' m^0(\alpha) = m_0'[G].$$

We have taken care to arrange the investigation of the

consequences of the equation IV, (3.3) in such a way that each step carries over to the inequality (2.2). The limit equations

$$m^0[G;\alpha] = O(1) \quad (\text{I fin.}), \qquad = o(R) \quad (\text{I infin.})$$

holding uniformly with respect to α entail the corresponding equations for the "little" term m'_0 in (2.2),

$$(2.3) \qquad m'_0[G] = (o),$$

which makes m'_0 negligible besides T. (For problems of rotational symmetry m'_0 is a constant which does not depend on G.)

In evaluating (2.1) we assume F and ρ to be homogeneous in the entire α-space,

$$F(t\alpha_1,\ldots,t\alpha_n) = F(\alpha_1,\ldots,\alpha_n) \qquad \text{for any } t > 0.$$

We may then replace (2.1) by the quotient of the two integrals

$$\int F(\alpha)\rho(\alpha)\cdot e^{-|\alpha_1|^2-\ldots-|\alpha_n|^2}\cdot d\alpha_1\, d\bar{\alpha}_1 \ldots d\alpha_n\, d\bar{\alpha}_n,$$

$$\int \rho(\alpha)\cdot e^{-|\alpha_1|^2-\ldots-|\alpha_n|^2}\cdot d\alpha_1\, d\bar{\alpha}_1 \ldots d\alpha_n\, d\bar{\alpha}_n = \pi^n A.$$

Evaluation of the first integral for F = N follows the same line of procedure as for the ordinary average. For two given vectors $x \neq 0$ and x' consider the plane $(\alpha x) = 0$ in the α-space, denote its area element by $d\sigma$ and form

$$(2.4) \quad S'(x,x') = \frac{2}{\pi^{n-1}}\cdot\int_{(\alpha x) = 0} \frac{|(x'\alpha)|^2}{|x|^2}\, e^{-|\alpha|^2}\rho(\alpha)\cdot d\sigma.$$

If one interprets x_i as the coordinates of our analytic curve and x'_i as their derivatives $\frac{dx_i}{dz}$ with respect to a local parameter z, the expression (2.4) which we then denote by S'_z is independent of the gauge factor and reacts like a density under transformation of the local parameter z. Our procedure leads to the general formula

$$\mathfrak{M}'N(\alpha) = \frac{1}{2\pi A}\cdot\int \phi S',$$

and thus the inequality (2.2) turns into

$$(2.5) \qquad \frac{1}{2\pi}\int \phi S' \leqq A(T + m_0').$$

The constant A is not only independent of G, but also of the curve $\mathfrak{C}$.

The left side of (2.5) equals

$$c'R + \int_0^R (R - r)Q'(r)\,dr$$

with

$$c' = \frac{1}{2\pi}\int_{K_0} S', \qquad Q'(r) = \frac{1}{2\pi}\int_{\Gamma_r} S_\vartheta' d\vartheta,$$

and as

$$\exp\left\{\frac{1}{2\pi}\int_{\Gamma_r} \log S_\vartheta'\cdot d\vartheta\right\} \leqq \frac{1}{2\pi}\int_{\Gamma_r} S_\vartheta' d\vartheta,$$

we obtain for

$$(2.6) \qquad 2\Theta(r) = \frac{1}{2\pi}\int_{\Gamma_r} \log S_\vartheta'\cdot d\vartheta$$

the inequality

$$c'R + \int_0^R (R - r)e^{2\Theta(r)}dr \leqq A(T + m_0').$$

Add 1 + (1-c')R and observe that for sufficiently large G

$$1 + (1-c')R + Am_0' \leqq \begin{cases} T \\ \text{Const.} \end{cases}$$

in case of zero and positive capacity respectively. In terms of the symbol ω the resulting relation may be expressed as

$$(2.7) \qquad 2\Theta = \omega(T) - \omega_B(T)$$

where the parameter B can be chosen as A + 1 for zero capacity, and as any number $>$ A for positive capacity. (Remember that in the latter case the assumption [∞] Chap. IV, §6, is made once for all.)

This is the general scheme which the exploitation of (2.5) will have to follow. We need not attempt to prove the existence of the integral (2.6) in this generality. We shall presently specialize ρ so as to obtain for Θ a sum of quantities Ω_p and m_p, and under these special circumstances there will be no doubt of the existence of Θ nor of the legitimacy of the entire procedure. Finding it hard to hit on the right choice of ρ right away, we follow the historical development by first studying the lowest case n = 2, i. e. the meromorphic functions on $\mathfrak{R}$.

§3. Defect relation for meromorphic functions on a Riemann surface

Assume $\rho(\alpha)$ to be homogeneous in the stricter sense that the equation

$$\rho(\tau\alpha_1,\ldots,\tau\alpha_n) = \rho(\alpha_1,\ldots,\alpha_n)$$

holds for arbitrary complex $\tau \neq 0$. Since for n = 2

$$\alpha_1 x_1 + \alpha_2 x_2 = 0 \quad \text{entails} \quad \alpha_1 : \alpha_2 = x_2 : -x_1,$$

(2.4) changes into

$$(3.1)\qquad \begin{aligned} S'_z &= \rho(x_2,-x_1)\cdot\frac{2}{\pi}\int_{(\alpha x)=0}\frac{|(x'\alpha)|^2}{|x|^2}e^{-|\alpha|^2}d\sigma \\ &= \rho(x_2,-x_1)\cdot S_z. \end{aligned}$$

Let us now take a positive exponent $\lambda < 1$ and any number q of distinct points $b = b_1,\ldots,b_q$:

$$b_{j1}b_{k2} - b_{k1}b_{j2} \neq 0 \qquad (\text{for } k \neq j;\ k,j = 1,\ldots,q).$$

We form the product

(3.2) $$\rho(\alpha) = \rho(\alpha_1, \alpha_2) = (\prod \|b\alpha\|^2)^{-\lambda}$$

extending over the q points b. The q planes $(\alpha b) = 0$ in α-space are singularities for this weight function, but as we shall presently see, the condition $\lambda < 1$ secures convergence of the integral of $\rho(\alpha)$ over the unit sphere. One after the other the following formulas result:

$$\rho(x_2, -x_1) = (\prod \|b:x\|^2)^{-\lambda},$$

$$\log S_\vartheta^* = \log S_\vartheta + 2\lambda \sum \log \frac{1}{\|b:x\|},$$

$$\Theta(r) = \Omega(r) + \lambda \sum_b \tilde{m}(r;b) \quad \text{with}$$

$$\tilde{m}(r;b) = \frac{1}{2\pi} \int_{\Gamma_r} \log \frac{1}{\|b:x\|} \cdot d\vartheta.$$

The resulting relation

(3.3) $$\Omega + \lambda \sum_b \tilde{m}(b) = \tfrac{1}{2}\omega(T)$$

improves our former equation $\Omega = \frac{1}{2}\omega(T)$ by the additional positive sum on the left side, and the essential step toward the Nevanlinna defect relation as announced in §1, is done.

Substituting (3.3) into the formula of the second main theorem

$$V - 2T = (\Omega - \Omega^0) + \eta,$$

we find

(3.4) $$V + \lambda \sum \tilde{m}(b) = 2T + \eta + (o) + \tfrac{1}{2}\omega(T).$$

If we are again content with the rough estimates

$$\log T = o(T), \qquad (o) = o(T)$$

and <u>assume hypothesis</u> $\mathfrak{H}$, then the inequality

(3.5) $\|$ $$V + \lambda \sum \tilde{m}(b) < k'T$$

follows as one holding for almost all G, whatever the constant $k' > 2$. Set $k = k'/\lambda$ and observe that one may so dispose of $k' > 2$ and $\lambda < 1$ as to assign to k an arbitrary value > 2. We thus have proved:

THEOREM. Assume the order T of the analytic curve to fulfill the hypothesis $\mathfrak{H}$, and k to be any number > 2 and let b range over any number q of distinct points. Then the inequality

$$(3.6) \qquad \| \qquad V + \sum \tilde{m}(b) < kT$$

holds for almost all G.

In retrospect the reader will now understand what prompted us to make the above choice* (3.2) for $\rho(\alpha)$. Set

$$\underline{\lim} \frac{V}{T} = \sigma, \qquad \underline{\lim} \frac{\tilde{m}(b)}{T} = \delta(b) \qquad (G \rightarrow \mathfrak{R}).$$

Then

$$(3.7) \qquad \sigma + \sum_b \delta(b) \leqq 2.$$

In the hypothesis $\mathfrak{H}$ of the theorem and in these inferior limits, one may restrict oneself to <u>a given complete system</u> $\mathfrak{G}$ of regions G.

Since

$$\tilde{N}(b) + \tilde{m}(b) = T + \tilde{m}^0(b)$$

---

*When F. Nevanlinna (6th Congr. Math. Scand. Kopenhagen 1925, pp. 97-107) first conceived the method of weighted averages, he took for $\rho(\alpha)\cdot d\omega_\alpha$ the non-Euclidean area of $d\omega_\alpha$ which results from mapping one-to-one conformally upon the unit circle the universal covering surface of the α-sphere pricked in the q points $(b_1\alpha) = 0$. This choice is indeed the best possible and is closely related to Picard's own use of the modular function for his theorem. However, L. Ahlfors [8] observed that the nature of the singularities at the q points is the only feature which is of vital importance, and thus substituted the above elementary weight function for F. Nevanlinna's highly transcendental ρ.

(first main theorem) and $\tilde{m}^0(b) = o(T)$ the defect δ can also be defined by

$$1 - \delta(b) = \overline{\lim} \frac{\tilde{N}(b)}{T} .$$

(3.7) is Nevanlinna's defect relation, generalized from the z-plane to an arbitrary Riemann surface. Instead of the analytic curve in 2-space with the coordinates x_1, x_2, one can speak of the meromorphic function $f(\mathfrak{p}) = x_1/x_2$.

Picard's theorem is an immediate consequence. For if none of the q forms $x_1 b_{j2} - x_2 b_{j1}$ $(j = 1,\ldots,q)$ vanishes anywhere, then $\tilde{N}(b) = 0$, $\delta(b) = 1$ for $b = b_1,\ldots,b_q$, and the relation (3.7) is incompatible with $q > 2$. Hence Picard's theorem holds for a meromorphic function on a given Riemann surface provided its order satisfies the (restricted) hypothesis $\mathfrak{H}$.

To give an example for the application of (3.6) in which the V-part plays a role, consider a value $b = b_1/b_2$ such that no root of the equation $f(\mathfrak{p}) = b$ (no zero of $x_1 b_2 - x_2 b$) is simple. In the usual manner we convert $\mathfrak{R}$ into a covering surface over the w-sphere by the convention that a point $\mathfrak{p}$ of $\mathfrak{R}$ lies over the point w of the sphere if $f(\mathfrak{p}) = w$. This covering surface is the Riemann surface of the inverse function of f. Our hypothesis means that all points $\mathfrak{p}$ over $w = b_1/b_2$ are points of ramification; $b = b_1/b_2$ is then said to be a <u>value of total ramification</u>. We maintain:

> Under the hypothesis $\mathfrak{H}$ there can be no more than 4 values of total ramification.

If $x_1 b_2 - x_2 b_1$ vanishes ν times, $\nu \geq 2$, at a point $\mathfrak{p}_0$, then $x_1 x_2' - x_2 x_1'$ vanishes there $(\nu - 1)$-times. Since $\nu - 1 \geq \frac{1}{2}\nu$ for $\nu \geq 2$ we have

$$V \geq \tfrac{1}{2}\sum_b \tilde{N}(b)$$

if all q values b are values of total ramification. The relation (3.5) with $\lambda = 1/2$ then gives

$$\| \qquad \tfrac{1}{2}\sum \tilde{N}(b) + \tfrac{1}{2}\sum \tilde{m}(b) < k'T,$$

and since $\tilde{N}(b) + \tilde{m}(b) \geqq T$,

$$\| \qquad \tfrac{1}{2}qT \leqq k'T,$$

which leads to a contradiction unless $q \leqq 4$.

[[To what extent is the hypothesis $\mathfrak{H}$ essential for the validity of Picard's theorem? Let c be a positive number and suppose that a meromorphic function $f(\mathfrak{p})$ on a Riemann surface $\mathfrak{R}$ of zero capacity satisfies the inequality

$$(3.8) \quad \| \qquad \eta < cT$$

for almost all G (of a complete system). Then the defect sum cannot exceed $2 + c$ and thus Picard's theorem still holds, provided $c < 1$. If s is an integer and $c < s - 1$ then $f(\mathfrak{p})$ cannot omit more than s values. In case of positive capacity the left member of (3.8) is to be replaced by

$$\eta + L_\kappa = \eta + \frac{1+\kappa}{2} \log \frac{1}{I-R} \qquad (\kappa > 1).$$

But by some simple prestidigitation one may eliminate the factor $\frac{1+\kappa}{2}$. Indeed, according to (IV, 7.10)

$$(3.9) \quad \| \qquad \eta + \log \frac{1}{I-R} < cT$$

would imply

$$\| \qquad \frac{1}{2} \log \frac{1}{I-R} < cT$$

as soon as R is sufficiently near I, and multiplying the last inequality by $\kappa - 1$ and adding it to (3.9) one obtains

$$\| \qquad \eta + \frac{1+\kappa}{2} \log \frac{1}{I-R} < \kappa c \cdot T$$

Given $c < s - 1$ one may choose $\kappa > 1$ so that κc is still less than $s - 1$. Hence if

$$(3.10) \quad \| \qquad \eta + \{\log \frac{1}{I-R}\} < (s - 1 - \epsilon)T$$

holds for some positive constant ϵ and almost all G of a given complete system $\mathfrak{G}$, then the non-constant meromorphic function $f(\mathfrak{p})$ of order T cannot omit more than $s(\geqq 2)$ values. The braces on the left indicate that the term surrounded by them is present in case of positive capacity only. Picard's theorem ($s = 2$) is obtained under the condition

$$(3.11) \qquad \eta + \{\log \frac{1}{I-R}\} < (1 - \epsilon)T.$$

The following two examples show that this condition may not be replaced by one appreciably weaker; they will thus clarify the role of the double-barreled hypothesis $\mathfrak{H}$ for Picard's theorem.

(1) The function $f(z) = (1 - e^z)^{1/2}$ is single-valued on a Riemann surface $\mathfrak{R}$ the two sheets of which spread over the z-plane with the ramification points $z = 2\pi i n$ $(n = 0, \pm 1, \pm 2, \ldots)$. The function omits the 3 values $+1, -1, \infty$. Circles of radii r_0 and $R(>r_0)$ around the origin are used as inner conductor K_0 and exhausting region G respectively. Since

$$\phi(z) = \frac{1}{2} \log \frac{R}{|z|} \qquad \text{for } r_0 \leqq |z| \leqq R,$$

the sum $2\sum \phi(\mathfrak{w})$ extending over the critical points $\mathfrak{w}$ in $G - K_0$ is the valence in the z-plane of those zeros of $f^2 = 1 - e^z$ which lie outside the circle K_0 and hence, provided $r_0 < 2\pi$,

$$2\sum \phi(\mathfrak{w}) \sim \frac{R}{\pi} - \log \frac{R}{r_0},$$

$$\eta = \sum \phi(\mathfrak{w}) - \frac{1}{2} \log \frac{R}{r_0} \sim \frac{R}{2\pi} - \log \frac{R}{r_0}.$$

On the other hand

$$T\{f\} \sim \frac{1}{2}T\{f^2\} \sim \frac{R}{2\pi}.$$

In a similar fashion the algebroid function $f(z) = (1 - e^z)^{1/s}$, which omits $s + 1$ values, illustrates the fact that the condition (3.10) is the sharpest of its kind, as far as the part η of its left member is concerned.

(2) In order to cover the $\{ \}$-term in (3.11) and (3.10), consider the unit circle $|\omega| < 1$ in a complex ω-plane as our Riemann surface $\mathfrak{R}$ and the modular function $t(\omega)$ as our meromorphic function f on $\mathfrak{R}$. It omits the 3 values $0, 1, \infty$. (As in §1, the upper half-plane has been transformed into the unit circle.) With the same choice for K_0 and G as in the first example ($r_0 < R < 1$) the voltage of $G - K_0$ is $u = \log \frac{R}{r_0}$ and hence tends to $I = \log \frac{1}{r_0}$ with $G \longrightarrow \mathfrak{R}$. It is not difficult to prove that the order T(R) of the modular function behaves asymptotically for $R \longrightarrow 1$ exactly as

$$\log \frac{1}{1-R} \sim \log \frac{1}{I-u}.$$

More generally, the automorphic function $t_s(\omega)$ which maps the unit circle in the ω-plane one-to-one conformally upon the universal covering surface of the t-sphere pricked at $s + 1$ given points $t = t_1, \ldots, t_{s+1}$ omits these $s + 1$ values, and its order T(R) behaves asymptotically as $\frac{1}{s-1} \cdot \log \frac{1}{1-R}$. Cf. R. Nevanlinna [10], pp. 254 and 259.]]*

*Gunnar af Hällström in an important paper, Acta Ac. Aboensis 12, No. 8, 1940, of whose existence I learned only after completion of the manuscript of this Study, deals with meromorphic functions for multiply connected regions in the z-plane for which a sort of limiting potential ϕ exists. Both this and E. Ullrich's investigation of algebroid functions quoted in Chap. IV, §8, point toward a general theory of meromorphic functions on Riemann surfaces as developed here; the quantity η is clearly in evidence. Hällström's paper is a source of highly inte-

§4. Convergence of an integral

We have still to prove the integrability of the weight function (3.2). To this end we make use of the triangle property of distances in 2-space.

LEMMA 4. A. $\|a:x\| + \|b:x\| \geq \|a:b\|$ $(n = 2)$.

Proof. Supposing $a_1b_2 - a_2b_1 \neq 0$ we may set

$$x_i = sa_i + tb_i \qquad (i = 1,2).$$

Then

$$|a_1x_2 - a_2x_1| = |a_1b_2 - a_2b_1|\cdot|s|,$$

$$|b_1x_2 - b_2x_1| = |a_1b_2 - a_2b_1|\cdot|t|.$$

With the normalization $|a| = |b| = 1$ we find

$$|x|^2 = |s|^2 + |t|^2 + (a|b)s\bar{t} + (b|a)\bar{s}t,$$

and it remains to prove the inequality

$$(|s| + |t|)^2 \geq |s|^2 + |t|^2 + (a|b)s\bar{t} + (b|a)\bar{s}t,$$

which, however, follows at once from the trivial relations

$$|(a|b)s\bar{t}| = |(b|a)\bar{s}t| \leq |s|\cdot|t|.$$

The lemma will be used in the form

$$\|a\xi\| + \|b\xi\| \geq \|a:b\|. \tag{4.1}$$

Let now b run over a set $\sum$ of q distinct points b_1,

resting examples; e. g. there is one in which T is of lower order of magnitude than η. -- A different approach to Picard's theorem, leading to a very significant generalization but away from the analytic curves, is L. Ahlfors's metric-topological study of the Riemann surface over the w-sphere of the inverse of a meromorphic function $w = f(z)$. See L. Ahlfors [9], and for multiply-connected surfaces, J. Dufresnoy, C. R. Acad. Sci. Paris, 212, 1941, 746-749.

... , b_q and 2d be the least of the mutual distances $\| b_j:b_k \|$ $(j \neq k)$. According to (4.1), all q distances $\| b_j\alpha \|$ of the b_j from an arbitrary α, except perhaps one, are greater than or equal to d. Denote by b_1 this exceptional b if it exists; if there is no exceptional b choose b_1 at random among the b_j; then

$$\sum_j \log(\|b_j\alpha\|^{-2\lambda}) \leqq \log(\|b_1\alpha\|^{-2\lambda}) + 2\lambda(q-1) \log d^{-1}$$
$$\leqq \log \sum_j \|b_j\alpha\|^{-2\lambda} + 2(q-1) \log d^{-1}.$$

The exceptional b_1 varies with α, but if we skip the intermediate member, our inequality holds for every α and therefore

$$(4.2) \qquad (\prod_b \|b\alpha\|^2)^{-\lambda} \leqq (\tfrac{1}{d})^{2(q-1)} \cdot \sum_b \|b\alpha\|^{-2\lambda}:$$

<u>the product can be appraised in terms of the sum.</u>

Thus the integral of (3.2) will certainly converge if the integral of the individual term $\|b\alpha\|^{-2\lambda}$ does. In examining this we may take advantage of unitary invariance, and assume $b_1 = 1$, $b_2 = 0$. Then Lemma III, 2. B, becomes applicable with $n = 2$, $g(t) = t^{-\lambda}$, yielding the value

$$\int \|b\alpha\|^{-2\lambda} \cdot e^{-|\alpha|^2} d\alpha_1\, d\bar{\alpha}_1\, d\alpha_2\, d\bar{\alpha}_2 = \pi^2 \int_0^1 t^{-\lambda} dt = \frac{\pi^2}{1-\lambda}.$$

One now sees why the assumption $\lambda < 1$ is essential. As the constant A in (2.5) we may choose

$$A = \frac{q}{1-\lambda} (\tfrac{1}{d})^{2(q-1)}.$$

It depends on the exponent λ and the configuration of the points b, but on nothing else.

§5. Starting the investigation for arbitrary n

The experience gathered for $n = 2$ encourages us to try our luck with a weight function of the simple type

$$\rho(\alpha) = g(\|b\alpha\|^2)$$

where g(t) is a positive function in the interval $0 \leqq t \leqq 1$ and b any non-vanishing covariant vector which we normalize by $|b| = 1$. In order to compute S' for given vectors $x \neq 0$ and x' we adapt the unitary coordinate system to the data according to the scheme

$$(5.1) \quad \begin{cases} x = (x_1, \ 0, \ 0, \ \ldots, \ 0), \\ b = (b_1, \ b_2, \ 0, \ \ldots, \ 0), \\ x' = (x_2', \ x_2', \ x_3', \ \ldots, \ x_n'). \end{cases}$$

Then

$$(5.2) \quad \begin{aligned} S_z' = \frac{2}{\pi^{n-1}} \int & \frac{|x_2'\alpha_2 + \ldots + x_n'\alpha_n|^2}{|x_1|^2} \cdot e^{-|\alpha_2|^2 - \ldots - |\alpha_n|^2} \\ & \times g\left(\frac{|b_2|^2|\alpha_2|^2}{|\alpha_2|^2 + \ldots + |\alpha_n|^2}\right) d\alpha_2 \, d\bar{\alpha}_2 \ldots d\alpha_n \, d\bar{\alpha}_n . \end{aligned}$$

Develop the integrand

$$|x_2'\alpha_2 + \ldots + x_n'\alpha_n|^2 = \sum x_i'\bar{x}_j'\alpha_i\bar{\alpha}_j \quad (i, j = 2, \ldots, n).$$

As follows from the introduction of polar coordinates for the complex variables $\alpha_2, \ldots, \alpha_n$ the contributions of the mixed terms $i \neq j$ to the integral are zero, while each principal term $i = j$ contributes a positive amount. Retaining among the latter the term $i = j = 2$ only we thus find that the inequality (2.5) holds <u>a fortiori</u> if (5.2) is replaced by the smaller quantity

$$\begin{aligned} S_z' = \frac{2}{\pi^{n-1}} \int & \frac{|x_2'|^2 |\alpha_2|^2}{|x_1|^2} \, e^{-|\alpha_2|^2 - \ldots - |\alpha_n|^2} \\ & \cdot g\left(\frac{|b_2|^2|\alpha_2|^2}{|\alpha_2|^2 + \ldots + |\alpha_n|^2}\right) d\alpha_2 \, d\bar{\alpha}_2 \ldots d\alpha_n \, d\bar{\alpha}_n . \end{aligned}$$

After introduction of the function

$$(5.3)\quad \psi(s) = \int_0^\infty \cdots \int_0^\infty g\left(\frac{st_2}{t_2+\cdots+t_n}\right)\cdot t_2 e^{-t_2-\cdots-t_n} dt_2 \cdots dt_n$$

we can write

$$(5.4)\qquad \tfrac{1}{2}S_z' = \frac{|x_2'|^2}{|x_1|^2}\cdot\psi(|b_2|^2).$$

g is subject to the condition that

$$\int g(\|b\alpha\|^2)\cdot e^{-|\alpha|^2} d\alpha_1 d\bar{\alpha}_1 \ \cdots\ d\alpha_n d\bar{\alpha}_n = \pi^n e_0$$

be finite. In evaluating this integral we may assume $b = (1, 0, \ldots, 0)$. Then the condition of convergence assumes the form

$$(5.5)\quad \int_0^\infty \cdots \int_0^\infty g\left(\frac{t_1}{t_1+\cdots+t_n}\right)\cdot e^{-t_1-\cdots-t_n} dt_1 \cdots dt_n = e_0,$$

or by Lemma III, 2. B,

$$(5.6)\qquad (n-1)\int_0^1 g(t)\cdot(1-t)^{n-2} dt = e_0.$$

For the function $\psi(s)$ this has the consequence that its integral over the interval $(0,1)$ has the finite value e_0,

$$(5.7)\qquad \int_0^1 \psi(s) ds = e_0.$$

Indeed

$$\int_0^1 \psi(s) ds = \int_0^1 \int_0^\infty \cdots \int_0^\infty g\left(\frac{st_2}{t_2+\cdots+t_n}\right)\cdot t_2 e^{-t_2-\cdots-t_n} dt_2 \cdots dt_n ds$$

changes by the substitution

$$st_2 = \tau_1, \qquad (1-s)t_2 = \tau_2, \qquad t_3 = \tau_3, \ \ldots\ , t_n = \tau_n,$$

$$(t_2\, dt_2\, ds = d\tau_1\, d\tau_2)$$

into

$$\int_0^\infty \cdots \int_0^\infty g\left(\frac{\tau_1}{\tau_1+\ldots+\tau_n}\right)e^{-(\tau_1+\ldots+\tau_n)}d\tau_1 \ldots d\tau_n.$$

Is the positive function $\psi(s)$ subject to no other restriction than the convergence of the integral (5.7)? This is unlikely; for even if the integral equation (5.3) has a solution g, that solution might not be positive. We must clear up this matter before proceeding further. For $n = 2$ there is no problem, $g(s) = \psi(s)$. For $n \geq 3$ we first bring the relation between g and ψ into a simpler form by the same type of substitution which carried (5.5) into (5.6), namely

$$t_2 = \tau(s_3+\ldots+s_n), \qquad t_3 = (1-\tau)s_3,\ldots, \qquad t_n = (1-\tau)s_n$$

$$(t_2+\ldots+t_n = s_3+\ldots+s_n)$$

with the Jacobian

$$(s_3+\ldots+s_n)\ (1-\tau)^{n-3};$$

cf. the proof of Lemma III, 2. B. Thus $\psi(s)$ is changed into the product of

$$\int_0^1 g(s\tau)\cdot\tau(1-\tau)^{n-3}d\tau$$

by

$$\int_0^\infty \cdots \int_0^\infty (s_3+\ldots+s_n)^2\cdot e^{-s_3-\ldots-s_n}ds_3\ldots ds_n.$$

Since

$$\int_0^1 e^{-s}ds = 1, \qquad \int_0^1 se^{-s}ds = 1, \qquad \int_0^1 s^2e^{-s}ds = 2$$

and therefore

$$\int_0^\infty \cdots \int_0^\infty s_3(s_3+\ldots+s_n)e^{-s_3-\ldots-s_n}ds_3\ldots ds_n = n-1,$$

we find the value $(n-1)\cdot(n-2)$ for the second factor. Consequently

$$\psi(s) = (n-1)(n-2)\int_0^1 g(s\tau)\cdot\tau(1-\tau)^{n-3}d\tau,$$

$$s^{n-1}\psi(s) = (n-1)(n-2)\int_0^s tg(t)\cdot(s-t)^{n-3}dt.$$

In this form the equation is at once solvable,

$$s\cdot g(s) = \frac{1}{(n-1)!}\frac{d^{n-2}}{ds^{n-2}}\left(s^{n-1}\psi(s)\right)$$

and we see that $g(s)$ exists for every dimensionality n and is positive <u>if ψ and all its derivatives exist and are positive</u>. Thus $\psi(s) = (1-s)^{-\lambda}$ $(\lambda < 1)$ would be a permissible choice.

However, for the invariant expression of (5.4) it is convenient to have a factor $|b_2|^2$ in front of ψ. We therefore set

$$\psi(s) = s\cdot J(1-s)$$

so that

$$\text{(5.9)} \qquad sg(s) = \frac{1}{(n-1)!}\frac{d^{n-2}}{ds^{n-2}}\left(s^n\cdot J(1-s)\right).$$

Then

$$e_0 = \int_0^1 s\cdot J(1-s)ds = \int_0^1 (1-s)\cdot J(s)ds,$$

and we have to require first that

$$\text{(I)} \qquad \int_0^1 J(s)ds = e^*$$

converges, and secondly, that the steering function J and all its derivatives exist for $0 < s \leq 1$ and have alternating signs:

$$\text{(II)} \qquad J(s) > 0, \qquad \frac{dJ}{ds} < 0, \qquad \frac{d^2J}{ds^2} > 0, \ \ldots .$$

Next we turn to developing the invariant expression of (5.4) or

$$\frac{1}{2}S'_z = \frac{|x'_2\overline{b}_2|^2}{|x_1|^2}\cdot J(|b_1|^2).$$

x'_2 is found by projecting x' perpendicularly onto the 2-spread $\{x,b\}$, and thus we come upon the expression

$$(x|x)(x'|b) - (x'|x)(x|b),$$

which indeed for (5.1) reduces to $x_1\overline{x}_1\cdot x'_2\overline{b}_2$. Let us replace b by the contravariant vector $\beta = \underline{\mu}b$ with the components $\beta_i = \overline{b}_i$. Then we have

$$\frac{1}{2}S'_z = \frac{|(x|x)(x'\beta) - (x'|x)(x\beta)|^2}{|x|^6|\beta|^2}\cdot J(\|x\beta\|^2).$$

In this form the result holds irrespective of the normalization of β. The quantity

$$(x|x)(x'\beta) - (x'|x)(x\beta)$$

is the scalar product of

$$y = x(x'\beta) - x'(x\beta)$$

by x. But

$$|(y|x)|^2 = |y|^2\cdot|x|^2 - |[yx]|^2 \text{ and } [yx] = [xx'](x\beta).$$

Consequently

$$\frac{1}{2}S'_z = \frac{|x(x'\beta)-x'(x\beta)|^2}{|x|^4|\beta|^2}\cdot J(s) - \frac{|[xx']|^2}{|x|^4}\, sJ(s) = \frac{1}{2}S_z(\beta)-\frac{1}{2}S''_z$$

where $\|x\beta\|^2$ is to be substituted for s. We now introduce a further condition for the steering function, which is satisfied by $J(s) = s^{-\lambda}$ ($\lambda < 1$), namely

(III') $J^*(s) = s\cdot J(s) \leq 1$ (for $0 < s \leq 1$).

Then

$$S_z'' \leqq 2\cdot\frac{|[xx']|^2}{|x|^4} = S_z$$

and consequently

$$\frac{1}{2\pi}\int\phi S'' \leqq \frac{1}{2\pi}\int\phi S = T.$$

Thus we obtain the following formula at which we have been aiming throughout this section:*

$$\frac{1}{2\pi}\int\phi\cdot S(\beta) = \frac{1}{\pi}\int\phi\cdot\frac{|x(x'\beta)-x'(x\beta)|^2}{|x|^4|\beta|^2}\cdot J(\|x\beta\|^2)$$

(5.10)

$$\leqq e_1 T + (e_1-1)m_0'$$

with

$$e_1 = e_0 + 1 = \int_0^1 J(s)ds + \int_0^1 (1-sJ(s))ds.$$

For n = 2 and $J(s) = s^{-\lambda}$ this is practically the same formula as the one on which the proof of Nevanlinna's defect relation in §3 was based:

$$\frac{1}{\pi}\int\phi\frac{|[xx']|^2}{|x|^4}\cdot\|x\beta\|^{-2\lambda} \leqq e^*(T+m_0');$$

the only difference is that the constant factors of T and m_0' have been raised and lowered respectively: $e_1 > e^* > e_1 - 1$. The value of e_1 for $J(s) = s^{-\lambda}$ is

$$\frac{1}{1-\lambda} + \frac{1-\lambda}{2-\lambda} < \frac{1}{1-\lambda} + (1-\lambda).$$

Hence we may use the inequality (5.10) with $J(s) = s^{-\lambda}$ and

(5.11) $$e_1 = \frac{1}{1-\lambda} + (1-\lambda).$$

Let us restate the definition of m_0', which quantity had now better be designated as $m_0\{\beta\}$ because of its de-

*This formula (for $J(s) = s^{-\lambda}$ and the z-plane as Riemann surface) is due to Ahlfors, [12]. In broad outline we follow the same paper for the remainder of this Chapter, though there are many changes in detail.

pendence on β. It is the weighted average of

$$m^0(\alpha) = \frac{1}{2\pi}\int_\gamma \log \frac{1}{\|\alpha x\|}\cdot d\vartheta$$

over the α-sphere, the weight function being

$$g\left(\frac{|(\beta|\alpha)|^2}{|\beta|^2\cdot|\alpha|^2}\right). \tag{5.12}$$

Consider the case of zero capacity. Given any $\epsilon > 0$ we can ascertain a compact K^ϵ such that $m^0(\alpha) < \epsilon R$ for all α and all admissible $G \supset K^\epsilon$. Then the average m_0' of $m^0(\alpha)$ is less than ϵR for $G \supset K^\epsilon$, whatever the weight function ρ, in particular for every β in (5.12) and every steering function $J(s)$ satisfying the conditions (I) and (II). $g = g_n$ is connected with J by the relation (5.9). In case of positive capacity an inequality $m_0' < C$ holds for sufficiently large G, $G \supset K$, where C is independent of G and $\rho(\alpha)$ and the compact K is independent of $\rho(\alpha)$. Briefly, in both cases the relation $m_0\{\beta\} = (o)$ holds uniformly with respect to β and $J(s)$.

§6. Consequences for arbitrary rank

From the seed of the formula (5.10) there springs a bunch of relations similar to those developed from the equation (III,4.10) in Ch. III, §5. The first step consists in applying (5.10) to the dual curve $\mathfrak{C}^*$ with the coordinates ξ_i:

$$\frac{1}{\pi}\int\!\!\phi\, \frac{|\xi(\xi'b)-\xi'(\xi b)|^2}{|\xi|^2|(\xi b)|^2}\cdot J^*(\|\xi b\|^2) \leqq e_1T_{n-1}+(e_1-1)m^0_{n-1}\{b\} \tag{6.1}$$

where $m^0_{n-1}\{b\}$ is the weighted average of

$$\tilde{m}^0_{n-1}(a) = \frac{1}{2\pi}\int_\gamma \log\frac{1}{\|a\xi\|}\cdot d\vartheta = \frac{1}{2\pi}\int_\gamma \log\frac{1}{\|a:X^{n-1}\|}\cdot d\vartheta,$$

formed by means of the weight function

$$g_n\left(\frac{|(b|a)|^2}{|b|^2\cdot|a|^2}\right).$$

$*\xi$ equals X^{n-1}, $\|\xi b\| = \|b:X^{n-1}\|$. Moreover

$$(6.2) \qquad \xi(\xi' b) - \xi'(\xi b) = W\cdot *[bX^{n-2}]$$

where W is the Wronskian $*X^n$. Indeed, by definition of the star operation, the equation $[\xi\eta] = *[x_1,\ldots,x_{n-2}]$ implies

$$\begin{vmatrix} (u\xi), & (u\eta) \\ (b\xi), & (b\eta) \end{vmatrix} = |u,b,x_1,\ldots,x_{n-2}|,$$

u denoting an indeterminate covariant vector; hence by the same definition

$$\xi(b\eta) - \eta(b\xi) = *[bx_1\ldots x_{n-2}].$$

In our case $[\xi\xi']$ is the dual of $W\cdot X^{n-2}$. Thus (6.2) is proved, and the formula (6.1) expressed in terms of the original curve $\mathfrak{C}$ reads

$$\frac{1}{\pi}\int\phi \frac{|[X^{n-2}b]|^2\cdot|X^n|^2}{|X^{n-1}|^2\cdot|[X^{n-1}b]|^2}\cdot J^*\left(\frac{|[X^{n-1}b]|^2}{|X^{n-1}|^2}\right) \leqq e_1T_{n-1}+(e_1-1)m^0_{n-1}\{b\}.$$
(6.3)

We must try to bring J(s) as close to 1/s, and hence $J^*(s)$ as close to 1, as possible. Ignoring the "perturbing" factor J^* in (6.3) we see that the density S^{n-1} in the old formula

$$\frac{1}{2\pi}\int\phi S^{n-1} \leqq T_{n-1}$$

has been replaced by

$$S^{n-1}\frac{\|X^{n-2}:b\|^2}{\|X^{n-1}:b\|^2}.$$

The denominator represents a gain, the numerator a loss; but the gain is greater than the loss because $\|X^{n-2}:b\| \geqq \|X^{n-1}:b\|$.

Next we have to apply (6.3) to the curve arising from $\mathfrak{C}$ by projection along a given h-element $\{A^h\}$ whereby it turns into a curve in (n-h)-space. Set n - h = p + 1, or if p is given beforehand (p = 1,...,n-1) set h = n - p - 1. We have to substitute

$$[A^h,X^{p-1}],\ [A^h,X^p],\ [A^h,X^{p+1}] \text{ for } X^{n-2},\ X^{n-1},\ X^n,$$

while b now denotes a vector of modulus 1 perpendicular to $\{A^h\}$. The resulting inequality

$$\frac{1}{\pi}\int \phi\cdot F(A^h) \leqq e_1 T_p(A^h) + (e_1 - 1)m_p^0\{b,A^h\}$$

involves the density

$$(6.4)\quad F(A^h) = \frac{|[A^hX^{p+1}]|^2\cdot|[bA^hX^{p-1}]|^2}{|[A^hX^p]|^2\cdot|[bA^hX^p]|^2}\cdot J^*\left(\frac{|[bA^hX^p]|^2}{|b|^2\cdot|[A^hX^p]|^2}\right),$$

and $m_p^0\{b,A^h\}$ is the weighted average over the b'-sphere in the (p+1)-space $\{\sim A^h\}$ of

$$\frac{1}{2\pi}\int_\gamma \log\left(\frac{|[A^hX^p]|}{|[b'A^hX^p]|}\right)\cdot d\vartheta = \tilde{m}_p^0([b'A^h]) - \tilde{m}_p^0(A^h),$$

formed by means of the weight

$$(6.5)\qquad g_{p+1}\left(\frac{|(b|b')|^2}{|b|^2\ |b'|^2}\right)$$

Apply the appraisal (IV, 6.5) to $[b'A^h]$ rather than A^h. The result is a relation $m_p^0\{b,A^h\} = (o)$ holding uniformly for all A^h, all normalized b perpendicular to $\{A^h\}$, and all steering functions J(s) satisfying conditions (I) and (II). We have thus obtained the inequality

$$(6.6)\qquad \frac{1}{\pi}\int \phi\cdot F(A^h) \leqq e_{n-p-h}(T_p(A^h) + (o))$$

for h = n - p - 1. We wish to free the result from the hypothesis that b is perpendicular to $\{A^h\}$. To that end let b be an arbitrary normalized vector and $\tilde{b}$ its component perpendicular to $\{A^h\}$, $|\tilde{b}| = |[bA^h]| \leqq 1$. Substitution of $\tilde{b}$ for b in the expression $F(A^h)$ changes nothing except that $|\tilde{b}|^2$ takes the place of the factor

$|b|^2 = 1$ in the denominator of the argument of J^*. At this juncture we replace (III') by the sharper condition:

(III) $J^*(1) = 1$, $J^*(s)$ monotone increasing for $0 < s \leq 1$.

Then $J^*\left(\frac{s}{|\tilde{b}|^2}\right) \geq J^*(s)$, and thus the formula (6.6) holds for (6.4) without modification even if the normalized b is not perpendicular to $\{A^h\}$. (Had we retained the term $m_p^0\{b,A^h\}$, the density (6.5) would have to be replaced by

$$g_{p+1}\left(\frac{|(b|b')|^2 \cdot |A^h|^2}{|[bA^h]|^2 \cdot |b'|^2}\right),$$

the average being taken over the b'-sphere in the space $\{\sim A^h\}$ as before.)

Uniformity with respect to A^h is important because of the next move which carries our inequality (6.6) down step by step from $h = n - p - 1$ to $h = 0$ by means of the averaging process $\mathfrak{M}_{A^h \supset A^{h-1}}$. Being given an (h-1)-element $\{A^{h-1}\}$ we introduce a variable vector a perpendicular to $\{A^{h-1}\}$ and then form the ordinary average over a in the space $\{\sim A^{h-1}\}$. The projection $\tilde{X}^p$ of X^p along $\{A^{h-1}\}$ is represented by $[A^{h-1}, X^p]$. We use the abbreviation $[bX^p] = Y^{p+1}$. Hence we have to compute in the space $\{\sim A^{h-1}\}$ that quantity which, when written down for the original space, equals

$$\mathfrak{M}_a\left\{\frac{|[aX^{p+1}]|^2 \cdot |[aY^p]|^2}{|[aX^p]|^2 \cdot |[aY^{p+1}]|^2} \cdot J^*\left(\frac{|[aY^{p+1}]|^2}{|[aX^p]|^2}\right)\right\} = \mathfrak{M}_a f(a).$$

As for any homogeneous function, the average $\mathfrak{M}_a$ of $f(a)$ is to be computed by means of the equation

(6.7) $\pi^n \cdot \mathfrak{M}_a f(a) = \int f(a) \cdot e^{-|a|^2} \cdot da_1\, d\bar{a}_1 \ldots da_n\, d\bar{a}_n$.

It would be hard to ascertain its exact value. Instead we pass to the logarithm, again making use of the concavity of the logarithmic function:

$$\log \mathfrak{M}_a f(a) \geqq \mathfrak{M}_a \log f(a).$$

We then have to evaluate four quantities of the type

$$(6.8)\quad \mathfrak{M}_a(\log|[aX]|^2) \qquad \text{for } X = X^p, X^{p+1}, Y^p, Y^{p+1})$$

and as a fifth average

$$(6.9)\qquad \mathfrak{M}_a \log J^*.$$

By Lemma III, 5. A we know the result for (6.8):

$$\mathfrak{M}_a \log|[aX^p]|^2 = \log|X^p|^2 + \mathfrak{M}_a \log \frac{|a_{p+1}|^2+\ldots+|a_n|^2}{|a_1|^2+\ldots+|a_n|^2}.$$

This shows at once that these four contributions unite to form the logarithm

$$\log \frac{|X^{p+1}|^2\cdot|Y^p|^2}{|X^p|^2\cdot|Y^{p+1}|^2}.$$

The same method of computation based on unitary invariance is applicable to the expression (6.9). The argument of the function J^* in that expression is

$$\frac{|[aY^{p+1}]|^2}{|[aX^p]|^2}$$

where $X^p = [x_1 \ldots x_p]$ is any special p-ad $\neq 0$ and $Y^{p+1} = [bX^p]$. In an appropriate unitary coordinate system the components of $x_1, \ldots, x_p$, b will present the familiar recursion scheme

$$\begin{aligned} x_1 &= (x_{11},\ 0,\quad 0, \ldots, 0) \\ &\cdot\quad\cdot\quad\cdot\quad\cdot\quad\cdot\quad\cdot\quad\cdot \\ x_p &= (x_{p1}, \ldots, x_{pp}, 0, \ldots, 0), \\ b &= (b_1, \ldots\ldots, b_{p+1}, \ldots, 0). \end{aligned}$$

Then the argument equals

$$\frac{|Y^{p+1}|^2}{|X^p|^2} \cdot \frac{|a_{p+2}|^2 + \ldots + |a_n|^2}{|a_{p+1}|^2 + \ldots + |a_n|^2} .$$

We are led to introduce a new hypothesis for the steering function, namely

(IV) $J(st) \geqq J(s) \cdot J(t)$ (for $0 < s, t \leqq 1$),

but we are safe to do so because this functional inequality is satisfied for $J(s) = s^{-\lambda}$. It carries over from J to J^*. Thus we find

$$\mathfrak{M}_a \log J^*\left(\frac{|[aY^{p+1}]|^2}{|[aX^p]|^2}\right) \geqq \log J^*\left(\frac{|Y^{p+1}|^2}{|X^p|^2}\right) + \tag{6.10}$$
$$+ \mathfrak{M}_a \log J^*\left(\frac{|a_{p+2}|^2 + \ldots + |a_n|^2}{|a_{p+1}|^2 + \ldots + |a_n|^2}\right).$$

Set

$$-\int_0^\infty \ldots \int_0^\infty \log J^*\left(\frac{t_2 + \ldots + t_n}{t_1 + \quad + t_n}\right) e^{-t_1 - \ldots - t_n} dt_1 \ldots dt_n = j_{n-1} \ (>0).$$

Then our definition (6.7) of $\mathfrak{M}_a$ at once shows that the additive constant in the right member of (6.10) equals $-j_{n-p-1}$ and the following inequalities have been proved:

$$\log \mathfrak{M}_a f(a) \geqq \mathfrak{M}_a \log f(a) \geqq \log\left\{\frac{|X^{p+1}|^2 \cdot |Y^p|^2}{|X^p|^2 \cdot |Y^{p+1}|^2} \cdot J^*\left(\frac{|Y^{p+1}|^2}{|X^p|^2}\right)\right\} - j_{n-p-1},$$

$$\mathfrak{M}_{A^h \supset A^{h-1}} F(A^h) \geqq F(A^{h-1}) \cdot e^{-j_{n-h-p}} .$$

Remembering that the averaging process $(A^h \supset A^{h-1})$ carries $T_p(A^h)$ into $T_p(A^{h-1})$ we have thus established the relation (6.6) for all $h \leqq n - p - 1$ provided we define

$$e_p = e_1 \cdot e^{j_1 + \ldots + j_{p-1}}.$$

At the moment we are interested in the last inequality $h = 0$ only:

$$(6.11) \quad \frac{1}{2\pi} \int \phi S^p . \frac{\|b:X^{p-1}\|^2}{\|b:X^p\|^2} J^*(\|b:X^p\|^2) \leqq e_{n-p}(T_p + (o)).$$

Notice that for given rank p the factor e_p depends on nothing but the steering function. Let us compute it for $J(s) = s^{-\lambda}$. Here

$$j_{n-1} = (1-\lambda)\left[\mathfrak{M}_a \log \frac{|a_1|^2 + \ldots + |a_n|^2}{|a_n|^2} - \mathfrak{M}_a \log \frac{|a_2|^2 + \ldots + |a_n|^2}{|a_n|^2}\right]$$

The first term in the brackets has been evaluated in Ch. III, §2, to be $\frac{1}{1} + \ldots + \frac{1}{n-1}$, and hence the second term equals $\frac{1}{1} + \ldots + \frac{1}{n-2}$. Result: $j_{n-1} = (1-\lambda) \cdot \frac{1}{n-1}$,

$$(6.12) \quad e_p = e_1 \cdot \exp\{(1-\lambda)(\tfrac{1}{1} + \ldots + \tfrac{1}{p-1})\}.$$

§7. Sum into product

From a single point b we pass to a finite set $\sum$ of points, $a = a_1, \ldots, a_q$. We write the inequality (6.11) down for each of them and sum over $\sum$. An argument similar to that of §4 will enable us to replace the sum

$$\sum_a \frac{\|a:X^{p-1}\|^2}{\|a:X^p\|^2} \cdot J^*(\|a:X^p\|^2)$$

by the corresponding product. Let us temporarily denote by $\{a\}$ the individual term of this sum. Through any p points we can draw a p-element. Hence this transmutation of sum into product will be possible provided there is no p-element containing more than p of the points of $\sum$, and the inequality by which this is effected will be of the type

$$\frac{1}{p}\sum_a \log\{a\} \leqq \log(\frac{1}{p}\sum_a \{a\}) + \text{Const.}$$

Our hypothesis of "general position" is always satisfied if $q \leqq p$. In the opposite case $q > p$ it can be decided by a finite number of trials: For any $p + 1$ among the points of $\sum$, $a', \ldots, a^{(p+1)}$, one examines whether $[a', \ldots, a^{(p+1)}] \neq 0$.

We can free ourselves from the assumption of general position by attaching a positive <u>weight</u> or mass $w(a)$ to each point a of $\sum$ and <u>requiring that no p-element carries a load exceeding</u> 1. More explicitly, the sum $\sum w(a)$ is supposed never to exceed 1 when extended over those points a of $\sum$ which lie on an arbitrary p-element. In case of general position we choose $w(a) = 1/p$. Let us replace the q quantities $\|a:X^{p-1}\|^2 J^*(\|a:X^p\|^2)$ by any q non-negative numbers

$$f_s = f(a_s) \leqq 1 \qquad (s = 1, \ldots, q).$$

LEMMA 7. A. Under the assumption that no p-element carries a load > 1 an inequality prevails

$$(7.1) \quad \sum w(a) \log \frac{f(a)}{\|a:X^p\|^2} \leqq \log(1 + \sum w(a) \frac{f(a)}{\|a:X^p\|^2}) + \text{Const.}$$

in which Const. is not only independent of the arbitrary p-element $\{X^p\}$, but also of the q numbers $f(a) \leqq 1$.

Proof. Let us again use the abbreviation $\{a\}$ for $\frac{f(a)}{\|a:X^p\|^2}$. The lemma is fairly obvious if the sum $w = \sum w(a)$ extending over all q points a does not exceed 1.

This is necessarily so if $q \leqq p$. We simply have to apply the principle of the concavity of the logarithm in the form

$$\sum_a w(a) \log\{a\} + (1-w) \log 1 \leqq \log((1-w) + \sum_a w(a)\{a\}).$$

(The dummy term log 1 with the weight 1-w has been introduced in order to bring the total weight up to 1.) In a way the general case will be reduced to this special situation.

We take a lead from the proof of eq. (4.2) where, for a given α, we singled out that b in $\sum$ for which $\|b\alpha\|$ takes on the least value. What we do here is to arrange the q numbers $\|a:X^p\|$ in increasing order. Let therefore $t_1, \dots, t_q$ be any permutation of $1, \dots, q$ and set $a_{t_1} = a_1^*, \dots, a_{t_q} = a_q^*$. Let us say that a p-element $\{X^p\}$ belongs to the cell $\mathfrak{Z}(t_1,\dots,t_q)$ if

$$\|a_1^*:X^p\| \leqq \|a_2^*:X^p\| \leqq \dots \leqq \|a_q^*:X^p\| .$$

Some of these q! cells might be empty, but each $\{X^p\}$ belongs to at least one cell. We think of $\{X^p\}$ as spanned by a normal basis,

$$(7.2) \quad X^p = [x_1,\dots,x_p], \quad (x_i|x_k) = \delta_{ik} \quad (i,k = 1,\dots,p).$$

Determine the last index l for which

$$w' = w(a_1^*) + \dots + w(a_l^*) \leqq 1.$$

Then either $l = q$ or $w(a_1^*) + \dots + w(a_{l+1}^*) > 1$. It is impossible that $\|a_{l+1}^*:X^p\| = 0$ for any $\{X^p\}$ in $\mathfrak{Z}(t_1,\dots,t_q)$, because this would imply that the $l + 1$ points $a_1^*, \dots, a_{l+1}^*$ of total weight > 1 are incident with $\{X^p\}$. Consequently $\|a_{l+1}^*:X^p\|$ has a positive lower bound d if X^p, (7.2), varies over $\mathfrak{Z}(t_1,\dots,t_q)$. Therefore

$$\{a_s^*\} \leqq \frac{1}{\|a_s^*:X^p\|^2} \leqq \frac{1}{d^2} \qquad \text{for } s > l$$

and thus

$$\sum_a w(a)\log\{a\} \leqq w(a_1^*)\log\{a_1^*\}+\ldots+w(a_l^*)\log\{a_l^*\}+2(w-w')\log\frac{1}{d}.$$

The sum of the first l terms on the right side when treated according to the special case discussed in advance turns out to be less than or equal to

$$\log((1-w') + w(a_1^*)\{a_1^*\} + \ldots + w(a_l^*)\{a_l^*\})$$
$$\leqq \log(1 + \sum_a w(a)\cdot\{a\}).$$

d depends on the permutation $t_1, \ldots, t_q$, $d = d(t_1,\ldots,t_q)$; should $\mathfrak{Z}(t_1,\ldots,t_q)$ be vacuous, we set $d(t_1,\ldots,t_q) = 1$. Designating by d_0 the least of the q! positive numbers $d(t_1,\ldots,t_q)$ we obtain (7.1) with $\text{Const.} = 2w\cdot\log\frac{1}{d_0}$.

The result is equivalent to the inequality

$$\exp \sum_a w(a)\cdot\log\{a\} = \prod_a \{a\}^{w(a)}$$

$$\text{(7.3)} \qquad \leqq C(1+\sum_a w(a)\cdot\{a\}).$$

where $C = d_0^{-2w}$ depends only on the configuration $\sum$ of the mass points with their masses w(a). Let now f(a) assume its original value $\|a:X^{p-1}\|^2\cdot J^*(\|a:X^p\|^2)$. After replacing the letter b by a, multiply (6.11) by w(a), sum over the q points a, add the inequality

$$\frac{1}{2\pi}\int \phi S^p = T_p \leqq T_p + (o)$$

and finally combine the resulting relation

$$\frac{1}{2\pi}\int \phi S^p(1 + \sum w(a)\cdot\{a\}) \leqq (1 + we_{n-p})(T_p + (o))$$

with (7.3). Setting

$$(7.4)\quad F = \exp \sum_a w(a)\cdot \log \left\{ \frac{\|a:X^{p-1}\|^2}{\|a:X^p\|^2} \cdot J^*(\|a:X^p\|^2) \right\},$$

we arrive at the inequality

$$(7.5)\qquad \frac{1}{2\pi}\int \phi S^p F \leqq C(1 + we_{n-p})(T_p + (o)).$$

§8. The Ω-m-relation for points

It has been described, at the end of §2, how to evaluate such an inequality. The ideal relation would arise were we free to set $J(s) = 1/s$, $J^*(s) = 1$. Then we should have

$$\frac{1}{2\pi}\int_{\Gamma_r} \log(S^p F)d\vartheta = 2\Pi_p(r) \qquad \text{with}$$

$$\Pi_p(r) = \Omega_p(r) + \sum_a w(a)\{\tilde{m}_p(r;a) - \tilde{m}_{p-1}(r;a)\},$$

and thus come near to proving the inequality

$$(8.1)\quad \Pi_p = \Omega_p + \sum w(a)\{\tilde{m}_p(a) - \tilde{m}_{p-1}(a)\} \leqq 0.$$

Its most characteristic features are, - first that the compensating terms $\tilde{m}_p(a)$ for (covariant) points and not the quantities $m_p(A^p)$ appear in this relation, and secondly that it is the difference of two consecutive $\tilde{m}$, $\tilde{m}_p - \tilde{m}_{p-1}$, which really matters. This is not so strange because that difference is still positive, cf. II, §7.

Unfortunately we cannot prove the unqualified relation (8.1), but in three steps we shall try to come as close to it as seems possible. Set

$$2\varepsilon_p(r;a) = \frac{1}{2\pi}\int_{\Gamma_r} \log J^{*-1}(\|a:X^p\|^2)\cdot d\vartheta \quad (\geqq 0).$$

There is no doubt that this integral converges for $r < R$

since $J(s) \geqq 1$, $1/J^*(s) \leqq 1/s$. We obtain for

$$\frac{1}{4\pi}\int_{\Gamma_r} \log(S^pF)\cdot d\vartheta$$

the value

$$\Theta_p(r) = \Pi_p(r) - \sum w(a)\varepsilon_p(r;a)$$

which differs from $\Pi_p(r)$ by a "perturbing term", and (7.5) leads in familiar manner to the relation

$$cR + \int_0^R (R - r)e^{2\Theta_p(r)}\,dr \leqq C(1 + we_{n-p})(T_p + (o))$$

in which

$$c = \frac{1}{2\pi}\int_{K_0} S^pF$$

is a positive constant (depending on $J(s)$, though not on G).

The simplest choice we can make is $J(s) = s^{-\lambda}$ $(\lambda < 1)$; in the course of our investigation we have taken care to check that this steering function satisfies all the conditions (I)-(IV). For it we find

$$\varepsilon_p(r;a) = (1 - \lambda)\tilde{m}_p(r;a) \leqq (1 - \lambda)\{T_p(r) + \tilde{m}_p^0(a)\}$$

and thus

$$\Pi_p - (1 - \lambda)(wT_p + \sum_a w(a)\tilde{m}_p^0(a)) = \frac{1}{2}\omega(T)$$

where $B = 1 + C(1 + we_{n-p})$ may be chosen as the parameter of ω. For sufficiently large G the sum $\sum_a w(a)\tilde{m}_p^0(a)$ will be less than T_p and hence the inequality

(8.2) || $$\Pi_p < \epsilon\cdot T_p + L_\kappa$$

will hold for almost all G with $\epsilon = (1 - \lambda)(w + 2)$. But we can determine λ so that ϵ assumes an arbitrarily preassigned positive value.

We get somewhat closer to the ideal relation by choosing any constant $\kappa > 1$ and then forming the steering function

$$(8.3)\qquad J(s) = \frac{1}{s}(1 + \kappa^{-1}\log\frac{1}{s})^{-\kappa}.$$

Its integral over the interval (01) converges and, as the substitution

$$(8.4)\qquad t = \kappa^{-1}\log\frac{1}{s}, \qquad s = e^{-\kappa t}$$

at once shows, has the value

$$\int_0^1 J(s)ds = \kappa\int_0^\infty \frac{dt}{(1+t)^\kappa} = \frac{\kappa}{\kappa-1}.$$

As we shall presently see, (8.3) fulfills also the other conditions (II)-(IV). With this steering function we find

$$\varepsilon_p(r;a) = \kappa\cdot\frac{1}{2\pi}\int_{\Gamma_r}\log(1 + \kappa^{-1}\log\frac{1}{s})\cdot d\vartheta \qquad (s = \|a:X^p\|^2)$$

$$\leqq \kappa\cdot\log\{\frac{1}{2\pi}\int_{\Gamma_r}(1 + \kappa^{-1}\log\frac{1}{s})d\vartheta\}$$

$$= \kappa\cdot\log(1+2\kappa^{-1}\tilde{m}_p(r;a)) \leqq \kappa\cdot\log\{(1+2\kappa^{-1}\tilde{m}_p^0(a)) + 2\kappa^{-1}T_p(r)\}$$

and thus

$$\Pi_p - \kappa\cdot\sum w(a)\cdot\log(1+2\kappa^{-1}\tilde{m}_p^0(a) + 2\kappa^{-1}T_p) = \frac{1}{2}\omega(T_p).$$

Since

$$1 + 2\kappa^{-1}\tilde{m}_p^0(a) < 2(1 - \kappa^{-1})T_p$$

for sufficiently large G, (8.2) has now been improved to

$$(8.5)\quad \| \qquad \Pi_p < \kappa\frac{w+1}{2}\log T_p + L_\kappa.$$

This is a substantial improvement because it brings the order of magnitude of the perturbing term down to the same level as $\omega(T_p)$.

Let us check the conditions (II)-(IV) for the steering function (8.3). Condition (III) is evident. The functional relation (IV) amounts to

$$(1 + \kappa^{-1}\log\frac{1}{s})(1 + \kappa^{-1}\log\frac{1}{t}) \geqq 1 + \kappa^{-1}(\log\frac{1}{s} + \log\frac{1}{t})$$

for s, t $\leqq$ 1 and is thus verified. It remains to prove that (8.3) and its successive derivatives have alternating signs. With the above substitution (8.4), J(s) becomes

$$e^{\kappa\tau}, \qquad \tau = t - \log(1 + t)$$

while

$$\frac{d}{ds} = -\frac{1}{\kappa}e^{\kappa t}\cdot\frac{d}{dt}.$$

Hence we have to show that the functions J_n defined by

$$J_0 = e^{\kappa\tau}, \qquad J_n = e^{\kappa t}\frac{dJ_{n-1}}{dt} \qquad (n = 1,2,\ldots)$$

are all positive for $t > 0$. We maintain that J_n is of the following form

$$J_n = e^{\kappa(nt+\tau)}\frac{P_n(t)}{(1+t)^n} = e^{\tau_n}\cdot P_n(t),$$

$$\tau_n = \kappa(n + 1)t - (\kappa + n)\log(1 + t),$$

where $P_n(t)$ is a polynomial of degree n with non-negative coefficients. Proof: Form the derivative

$$\frac{d\tau_n}{dt} = \kappa(n+1) - \frac{\kappa+n}{1+t} = \frac{n(\kappa-1) + \kappa(n+1)t}{1 + t}$$

and thus obtain the recursion formula

$$P_{n+1}(t) = \{n(\kappa-1) + \kappa(n+1)t\}P_n(t) + (1+t)\frac{dP_n}{dt} \quad (n=0,1,\ldots).$$

Our third and best approximation to the ideal (8.1) is based on the observation due to Ahlfors that the relation

(7.5) holds even if the steering function $J(s)$ depends on G. Following Ahlfors, we set $J(s) = s^{-\lambda}$ with an exponent $\lambda = \lambda[G] < 1$ which tends to 1 as G exhausts the Riemann surface. In anticipation of this idea we have been careful to establish our relations uniformly for all admissible steering functions, in particular, uniformly for all positive exponents $\lambda < 1$ in $J(s) = s^{-\lambda}$. By (5.11) and (6.12)

$$e_1 \sim \frac{1}{1-\lambda}, \qquad e_{n-p} \sim \frac{1}{1-\lambda}, \qquad \text{for } \lambda \longrightarrow 1 .$$

Going through the same steps as before and taking account of $(o) = o(T_p)$ we obtain the inequality

$$\int_0^R (R-r)\, e^{2\Theta_p(r)}\, dr \leqq (1 + \epsilon) Cw \cdot \frac{T_p}{1-\lambda} \qquad \text{for } G \supset K^{\epsilon}.$$

Here ϵ is any preassigned positive number, and K^{ϵ} a suitable compact part of $\mathfrak{R}$ depending on ϵ. However, there is the new feature that with λ depending on G the quantity

$$(8.6) \qquad \Theta_p(r) = \Pi_p(r) - (1-\lambda)\sum\nolimits_a w(a)\tilde{m}_p(r;a)$$

now is not a function of G_r alone, but of G and r separately because the perturbing term written in explicit form equals

$$(1-\lambda[G])\sum\nolimits_a w(a)\tilde{m}_p[G_r;a].$$

But

$$\tilde{m}_p[G_r;a] \leqq T_p[G_r] + \tilde{m}_p^0[G_r;a]$$

and as

$$T_p[G_r] \leqq T_p[G], \qquad \tilde{m}_p^0[G_r;a] = \tilde{m}_p^0[G;a]$$

we have

$$\sum\nolimits_a w(a)\tilde{m}_p[G_r;a] \leqq w \cdot T_p[G] + \sum\nolimits_a w(a)\tilde{m}_p^0[G;a].$$

Hence let us set

(8.7) $1 - \lambda[G] = \frac{1}{2}\{wT_p[G] + \sum_a w(a)\tilde{m}_p^0[G;a]\}^{-1}$.

Then

$$\exp\ \{2(1-\lambda)\sum w(a)\tilde{m}_p(r;a)\} \leqq e,$$

and the following inequality results

$$\int_0^R (R-r)e^{2\Pi_p(r)}dr \leqq (1+\epsilon)2eCwT_p\{wT_p + \sum w(a)\tilde{m}_p^0(a)\}.$$

$\Pi_p(r) = \Pi_p[G;r]$ is a function of G_r alone. Hence after adding $1 + R = o(T_p^2)$ on both sides, we have proved the following

THEOREM. Under the assumption that the weights $w(a)$ attached to the points a of a finite set $\sum$ are such that no p-element carries a load exceeding unity,

(8.8) $\Pi_p = \Omega_p + \sum_a w(a)(\tilde{m}_p(a) - \tilde{m}_{p-1}(a)) = \frac{1}{2}\omega(T_p^2)$,

the parameter of ω being any constant $> 2eCw^2$.

A function which is $\frac{1}{2}\omega(T_p^2)$ is less than $\kappa \log T_p + L_\kappa$ for almost all G. Hence of our three approximations Ahlfors's result is essentially the best. The choice (8.7) with the definite numerical factor $\frac{1}{2}$ is dictated by the fact that $\frac{e^x}{x}$ takes on its minimum for $x = 1$.

One could combine the second and third approximations by setting

$$J[G;s] = \frac{1}{s}(1 + l^{-1}\log\frac{1}{s})^{-\kappa}$$

where $\kappa > 1$ is a fixed constant, but $l = l[G]$ so depends on G as to tend to infinity with $G \to \mathfrak{R}$. We have convinced ourselves that the best choice of $l[G]$ gives the same result $\omega_B(T_p^2)$ with the same parameter values $B > 2eCw^2$ as before. This seems to indicate that Ahlfors's estimate is an optimum which cannot be improved by devising more refined steering functions J[G;s].

§9. Discussion of the defect relations for points

For points in "general position" we can choose w(a) = 1/p without violating the condition that no p-element carries a load exceeding unity. The inequality (8.2) then becomes

$$(9.1) \quad \| \quad \Omega_p + \frac{1}{p}\sum_a(\tilde{m}_p(a)-\tilde{m}_{p-1}(a)) < \epsilon T_p + L_\kappa,$$

in particular, for p = 1, under the sole assumption that the points are distinct,

$$(9.2) \quad \| \qquad \Omega_1 + \sum_a \tilde{m}_1(a) < \epsilon T_p + L_\kappa.$$

This is in the closest analogy to what we have found for analytic curves in 2-space (meromorphic functions).

Let us discuss the simplest analytic curves in the z-plane, namely the rational and the exponential curves of type (II,5.1). For the moment we return to designating the radius itself by r so that $z = r\cdot e^{i\vartheta}$. A rational curve has a point at infinity, $z = \infty$. Adjustment of coordinates for the corresponding branch of the curve gives the expansions

$$x_1 = t^{\delta_1} + \dots,\ x_2 = t^{\delta_2} + \dots,\ \dots \quad (0 = \delta_1 < \delta_2 < \dots)$$

in terms of the parameter $t = 1/z$. The point $z = \infty$ is then located at

$$(9.3) \qquad x_1 : x_2 : \dots : x_n = 1:0: \dots :0.$$

The product [ax] with a fixed vector $a \neq 0$ does not vanish at infinity unless the point a coincides with (9.3); in the latter case the vanishing order of the dyad [ax] is δ_2. Hence

$$\tilde{m}_1(r;a) \sim \begin{cases} 0 & \text{if } (a_1:\ldots:a_n) \neq (1:0:\ldots:0), \\ \delta_2 \log r & \text{in the opposite case.} \end{cases}$$

On the other hand, the (1,2)-component of $[x, \frac{dx}{d\vartheta}]$ vanishes to the order δ_2 and all other components to a higher order; consequently $\Omega_1(r) \sim -\delta_2 \log r$. These facts show that there is a mutual compensation between Ω_1 and the $\tilde{m}_1(a)$ in such manner that $\Omega_1 + \sum \tilde{m}_1(a)$, the sum extending over a finite set $\sum$ of distinct points a, is essentially negative, namely ~ 0 if (9.3) belongs to the set, $\sim -\delta_2 \log r$ if not. It was this remark, together with Nevanlinna's defect relation for meromorphic functions, which first suggested the formula (9.2) for meromorphic curves; H. and J. Weyl, [11].

Perhaps the example of the exponential curves is even more illustrative because for them the "defect" is not concentrated in one point. Let $(e_1, \ldots, e_n)$ be the vector basis in terms of which the curve is given by (II, 5.1). We resume the notations previously used in studying such "special" exponential curves. Moreover, let $q_1^{(i)}(\vartheta), \ldots, q_{n-1}^{(i)}(\vartheta)$ denote the numbers $P_1(\vartheta), \ldots, P_{i-1}(\vartheta), P_{i+1}(\vartheta), \ldots, P_n(\vartheta)$ arranged in descending order, $q_1^{(i)}(\vartheta) \geq \ldots \geq q_{n-1}^{(i)}(\vartheta)$. Then in analogy to II, (5.3),

$$\tilde{m}_1(r;e_k) \sim \frac{r}{2\pi} \int_0^{2\pi} \{q_1(\vartheta) - q_1^{(k)}(\vartheta)\} d\vartheta.$$

But $q_1^{(k)}(\vartheta) = q_1(\vartheta)$ unless P_k is the largest of all the P_i, and in that case $q_1^{(k)}(\vartheta) = q_2(\vartheta)$. Thus

$$\tilde{m}_1(r;e_k) \sim \frac{r}{2\pi} \int_{\Theta_k} (q_1(\vartheta) - q_2(\vartheta)) d\vartheta.$$

On the other hand

$$\Omega_1(r) \sim -\frac{r}{2\pi} \int_0^{2\pi} (q_1(\vartheta) - q_2(\vartheta)) d\vartheta$$

and therefore

$$\Omega_1(r) + \sum_{k=1}^{n} \tilde{m}_1(r;e_k) \sim 0.$$

The phenomenon of compensation could not reveal itself in a more striking manner. For any $a \neq e_1, \ldots, e_n$ the term $\tilde{m}_1(r;a)$ is clearly equivalent to zero.

This illustration covers also the relation (9.1) for arbitrary rank which was first discovered by Ahlfors and which is not so easy to verify for rational curves. Indeed

$$\tilde{m}_p(r;e_k) \sim \frac{r}{2\pi}\int_0^{2\pi} \{(q_1+\ldots+q_p) - (q_1^{(k)}+\ldots+q_p^{(k)})\} d\vartheta,$$

$$\tilde{m}_p(r;e_k) - \tilde{m}_{p-1}(r;e_k) \sim \frac{r}{2\pi}\int_0^{2\pi} (q_p(\vartheta) - q_p^{(k)}(\vartheta)) d\vartheta.$$

Now $q_p^{(k)}$ will be q_{p+1} if $P_k(\vartheta)$ is one of the first p of the P_i's, otherwise it will be q_p. Hence

$$\tilde{m}_p(r;e_k) - \tilde{m}_{p-1}(r;e_k) \sim \frac{r}{2\pi}\int (q_p(\vartheta) - q_{p+1}(\vartheta)) d\vartheta$$

with the integral extending over the set $\Theta_k^{(p)}$ of those ϑ for which $P_k(\vartheta) = q_1(\vartheta)$ or $P_k(\vartheta) = q_2(\vartheta)$ or ... or $P_k(\vartheta) = q_p(\vartheta)$. Clearly the join of the sets $\Theta_k^{(p)}$ for $k = 1, \ldots, n$ covers the whole periphery precisely p-times. Consequently

$$\frac{1}{p}\sum_k \{\tilde{m}_p(r;e_k) - \tilde{m}_{p-1}(r;e_k)\} \sim \frac{r}{2\pi}\int_0^{2\pi} (q_p(\vartheta) - q_{p+1}(\vartheta)) d\vartheta,$$

$$\Omega_p(r) + \frac{1}{p}\sum \{\tilde{m}_p(r;e_k) - \tilde{m}_{p-1}(r;e_k)\} \sim 0.$$

Not only does this example exhibit most convincingly the inner connection between Ω_p and the defect difference $\tilde{m}_p(a) - \tilde{m}_{p-1}(a)$, but it proves at the same time that our relations are the strongest of their kind.

We return to the general theory and as usual combine (9.1) with the formula of the Second Main Theorem, and assume during the remainder of this section that <u>the hypothesis $\mathfrak{H}$ is fulfilled</u>. Then we obtain, for a set $\sum$ of points a in general position and for almost all G,

$$(9.4)\quad \| \; V_p + (T_{p+1}+T_{p-1}) + \frac{1}{p}\sum\{\tilde{m}_p(a)-\tilde{m}_{p-1}(a)\} < (2+\epsilon)T_p.$$

The striking consequences which we drew from this relation for analytic curves in 2-space (meromorphic functions) arose from its combination with the First Main Theorem

$$\tilde{N}_1(a) + \tilde{m}_1(a) = T_1 + \tilde{m}_1^0(a) \qquad (n = 2),$$

which reveals the fact that $\tilde{m}_1(a)$ measures a defect of the valence $\tilde{N}_1(a)$ of the a-places of the function under consideration. For higher n we have instead (see II, §7)

$$\tilde{N}_1(a) + \tilde{m}_1(a) = \Lambda_1$$

where Λ_1 is not the order T_1 but its reduction by projection of the curve from the point a, and in general it is to be expected that this loss is small compared to T_1 itself. Hence $\tilde{m}_1(a)$ is, as it were, a <u>defect of a defect</u>. The same is true for $\tilde{m}_p(a)$, and not even $\tilde{m}_p(a)$ but the difference $\tilde{m}_p - \tilde{m}_{p-1}$ is the relevant quantity in the relation (9.4)!

The latter circumstance need not discourage us too much. If we replace p by another index p', then multiply by p' and sum over p' = 1, ..., p, we find (cf. III, §7)

$$(9.5)\quad \| \qquad pT_{p+1} + \sum_a \tilde{m}_p(a) < (p + 1 + \epsilon)T_p,$$

and thus a defect relation involving $\tilde{m}_p(a)$ only results. The V-part $1\cdot V_1 + \dots + p\cdot V_p$ in the left member has been omitted. The underlying hypothesis is that no element of rank $\leq$ p carries a load exceeding 1. If the number q of points a is greater than p, the conditions concerning the elements of rank lower than p are implied in the condition for p-elements. But in case $q \leq p$ they are not redundant, and the whole hypothesis amounts to assuming that the q points are linearly independent. We now interpret the phrase "general position" accordingly.

However it seems wise to look upon (9.4) or, still better without any hypothesis about "general position", upon the inequality

$$(9.6)\quad \| V_p+(T_{p+1}+T_{p-1})+\sum_a w(a)\{\tilde{m}_p(a)-m_{p-1}(a)\}<(2+\epsilon)T_p$$

as the real basic relation.

At this juncture we make a remark which in a systematic exposition should have been made long before; namely that we may apply all our results to the dual curve $\mathfrak{C}^*$ and then discuss what they mean for the original curve. We know that $T^*_{n-p} = T_p$. The distance $\|a:X^{n-p}\|$ becomes $\|\alpha:\Xi^{n-p}\|$. But $\{\Xi^{n-p}\} = \{*X^p\}$, and after setting $\alpha = \underline{\mu}a$ (i. e. $\alpha_i = \overline{a}_i$ in a normal coordinate system), by (I, 4.6)

$$\|\underline{\mu}a:*X^p\| = \|a:\sim X^p\|.$$

Project the vector a of length 1 along $\{X^p\}$ upon $\{\sim X^p\}$. Then $\|a:X^p\|$ is the magnitude of the <u>projection</u>, whereas $\|a:\sim X^p\|$ is the magnitude of the <u>perpendicular</u>. Hence the quantities $\tilde{m}_p(a)$, $\tilde{N}_p(a)$ for the projection change into the corresponding quantities for the perpendicular $\hat{m}_p(a)$, $\hat{N}_p(a)$, the latter being the valence in G of the zeros of $\|a:\sim X^p\|$ and

$$\hat{m}_p(a) = \frac{1}{2\pi}\int_\Gamma \log\frac{1}{\|a:\sim X^p\|}\cdot d\vartheta.$$

Equation (II, 7.3) (for h=1) turns into

$$\hat{N}_p(a) + \hat{m}_p(a) - \hat{m}^0_p(a) = T_p - T^*_{n-p}(\underline{\mu}^{-1}a).$$

The interpretation of $T^*_{n-p}(\underline{\mu}^{-1}a)$ in terms of the original curve is somewhat clumsy. Besides $\{\Xi^{n-p}\} = \{*X^p\}$ we have $\{\Xi^{n-p-1}\} = \{*X^{p+1}\}$. Hence the inequalities (6.11),

(7.5) remain valid if one replaces e_{n-p} on the right side by e_p and

$$\{a\} = \frac{\|a:X^{p-1}\|^2}{\|a:X^p\|^2}\cdot J^*(\|a:X^p\|^2) \text{ by } \frac{\|a:\sim X^{p+1}\|^2}{\|a:\sim X^p\|^2}\cdot J^*(\|a:\sim X^p\|^2).$$

It is this dual formula rather than the formula (6.11), which for $p = 1$ coincides with the basic inequality (5.10) from which we started. (8.1), (8.2) change into

$$\| \quad \Omega_p + \sum_a w(a)(\hat{m}_p(a) - \hat{m}_{p+1}(a)) < \epsilon T_p + L_\kappa,$$

valid under the hypothesis that no (n-p)-spread carries a load exceeding unity. For "general position", i. e. if no (n-p)-spread contains more than $n - p$ of the points a, the inequality reduces to

$$\| \quad \Omega_p + \frac{1}{n-p}\sum_a (\hat{m}_p(a) - \hat{m}_{p+1}(a)) < \epsilon T_p + L_\kappa.$$

Provided general position prevails in the stricter sense that no spread of the dimensionalities $n - p, \ldots, n - 1$ carries a load greater than 1, a procedure arising by the inversion $p \to n - p$ from that which yielded the relation (9.5) results in the corresponding inequality

$$\| \quad (n-p)T_{p-1} + \sum_a \hat{m}_p(a) < (n-p+1+\epsilon)T_p.$$

Let us summarize.

THEOREM. Under the hypothesis $\mathfrak{H}$ the following relations hold for almost all G:

$$(9.7) \quad \| \; V_p + (T_{p+1} + T_{p-1}) + \sum_a w(a)(\tilde{m}_p(a) - \tilde{m}_{p-1}(a)) < (2+\epsilon)T_p,$$

$$(9.7') \quad \| \; V_p + (T_{p+1} + T_{p-1}) + \sum_a w(a)(\hat{m}_p(a) - \hat{m}_{p+1}(a)) < (2+\epsilon)T_p,$$

$$(9.8) \quad \| \; (1\cdot V_1 + \ldots + pV_p) + pT_{p+1} + \sum_a \tilde{m}_p(a) < (p+1+\epsilon)T_p,$$

$$(9.8') \quad \| \{1\cdot V_{n-1}+\ldots+(n-p)\cdot V_p\} + (n-p)T_{n-p} + \sum_a \hat{m}_p(a) <$$
$$< (n-p+1+\epsilon)T_p.$$

In (9.7) the weights w(a) are such as to load no p-spread by more than unity, in (9.7') no (n-p)-spread is supposed to carry a higher load. (9.8) holds under the assumption that no p'-spread contains more than p' of the points a, for p' = 1, ..., p; (9.8') under the same assumption for p' = n-1, ..., n-p.

§10. Incidences of higher order. Formulation of the general defect relation

It is clear that the defect relations for a set of points ought to be generalized so as to cover a set $\sum$ of elements $\{A^h\}$ of arbitrary dimensionality h, $1 \leq h \leq n-1$. By a veritable *tour de force* Ahlfors reached this goal. We repeat his argument in a modified form and without his assumption of general position.

First we state a lemma for the ω-symbol. Henceforward we shall use ω in the special sense of ω_1 so that $\omega_B(T)$ is now written as $\omega(BT)$.

LEMMA 10. A. Let f_i and T_i be functions of G and the B_i positive constants (i = 1,...,m). Then the relations

$$f_1 = \omega(B_1T_1), \ \ldots, \ f_m = \omega(B_mT_m)$$

imply

$$\rho_1\cdot f_1 + \ldots + \rho_m\cdot f_m = \omega(\rho_1B_1T_1+\ldots+\rho_mB_mT_m)$$

if ρ_i are any positive constant weights of sum 1.

Proof. The lemma is an immediate consequence of the concavity of log in the form

$$\rho_1 \cdot f_1 + \dots + \rho_m \cdot f_m \leqq \log(\rho_1 \cdot e^{f_1} + \dots + \rho_m \cdot e^{f_m}),$$

$$e^{\rho_1 f_1 + \dots + \rho_m f_m} \leqq \rho_1 \cdot e^{f_1} + \dots + \rho_m \cdot e^{f_m},$$

or, what amounts to the same, of the general inequality between the geometric and arithmetic mean of the quantities $e^{f_1}, \dots, e^{f_m}$.

Let $\{A^h\}$, $\{X^p\}$ be any two elements of ranks h, p respectively, and i any of the numbers 0, 1, ..., h-1. The element $\{A^h\}$ is said to be i-incident with $\{X^p\}$ provided every $\{A^{h-i}\}$ of rank h - i contained in $\{A^h\}$ satisfies the relation $[A^{h-i}X^p] = 0$. Set $l = h-i$. If $A^h = [a_1 \dots a_h]$ it is sufficient to require this relation for all the $\binom{h}{l} = \binom{h}{i}$

(10.1) A^l of the form $[a_{j_1}, \dots, a_{j_l}]$ $\quad (j_1 < \dots < j_l)$.

The i-incidence could also be described by saying that any l vectors of $\{A^h\}$ (or any l among the vectors $a_1, \dots, a_h$) are linearly dependent modulo $\{X^p\}$. Ordinary incidence is 0-incidence. i-incidence implies (i-1)-incidence.

$\|A^h : X^p\|$ is a unitarily invariant quantity the vanishing of which indicates ordinary incidence. Is it possible to construct a unitarily invariant quantity whose vanishing indicates i-incidence? We choose a normal basis $a_1, \dots, a_h$ of $\{A^h\}$ and then form the sum

(10.2) $$\binom{h}{l} \cdot d_i^2 = \sum \|A^l : X^p\|^2$$

extending over the $\binom{h}{l} = \binom{h}{i}$ elements (10.1) Clearly $d_i \leqq 1$ and d_i vanishes if and only if $\{A^h\}$ is i-incident with $\{X^p\}$. We maintain that this quantity is unitarily invariant. Indeed, let $b_1, \dots, b_h$ be another normal basis of $\{A^h\}$ by means of which we form all the $B^l = [b_{j_1}, \ldots, b_{j_l}]$. We contend that

$$\sum \|B^l : X^p\|^2 = \sum \|A^l : X^p\|^2.$$

The unitary invariance of the scalar product

$$(X|Y) = \sum_j{}' X(j_1 \ldots j_l)\overline{Y}(j_1 \ldots j_l)$$

proves that a unitary transformation U in a space of h dimensions induces a unitary transformation $U^{(l)}$ in the $\binom{h}{l}$-dimensional space of all its l-ads. $b_1, \ldots, b_h$ arise from $a_1, \ldots, a_h$ by a unitary transformation U, and the B^l are connected with the A^l by the induced transformation $U^{(l)}$. The same is true for the quantities $[B^l X^p]$ and $[A^l X^p]$. Hence

$$\sum |[B^l X^p]|^2 = \sum |[A^l X^p]|$$

A further pleasant property of the d_i is laid down in the inequality

$$d_{i-1}^2 \leqq \binom{h}{i} d_i^2,$$

which illustrates quantitatively the fact that i-incidence is stronger than (i-1)-incidence. Consider the following situation.

Let $x(A^l)$ be $\binom{h}{l}$ positive numbers associated with the A^l, (10.1), which we derived from a fixed normal basis $a_1, \ldots, a_h$ of $\{A^h\}$, and let $y(A^{l-1})$ have the same significance for l - 1. We form

$$\sum \frac{x(A^l)}{y(A^{l-1})}, \tag{10.3}$$

the range of the sum being limited by the accessory condition $\{A^{l-1}\} \subset \{A^l\}$. More precisely, for any $A^l = [a_{j_1}, \ldots, a_{j_l}]$ the element $\{A^{l-1}\}$ runs over the l elements obtained by dropping one of the factors $a_{j_1}, \ldots, a_{j_l}$;

for any one of the $A^{l-1} = [a_{j_1},\ldots,a_{j_{l-1}}]$ the element $\{A^l\}$ runs over the h - l + 1 elements obtained by adding one more factor $a_j(j \neq j_1,\ldots,j_{l-1})$. The number of terms in the sum is

$$\binom{h}{l}\cdot l = \binom{h}{l-1}\cdot(h-l+1).$$

Form the product of this sum by $\sum y(A^{l-1})$. In computing it term by term, multiply for a given A^l each of the corresponding l terms $x(A^l)/y(A^{l-1})$ in (10.3) by the term $y(A^{l-1})$ of the second sum. The contribution of these l individual products amounts to $l\cdot x(A^l)$. Therefore

$$\sum \frac{x(A^l)}{y(A^{l-1})}\cdot\sum y(A^{l-1}) \geq l\cdot\sum x(A^l). \tag{10.4}$$

Set

$$\sum x(A^l) = \binom{h}{l}\cdot x, \qquad \sum y(A^{l-1}) = \binom{h}{l-1}\cdot y,$$
$$\sum \frac{x(A^l)}{y(A^{l-1})} = (h-l+1)\cdot c. \tag{10.5}$$

Then (10.4) becomes

$$\frac{x}{y} \leqq c. \tag{10.6}$$

Apply this relation to

$$x(A^l) = \|A^l:X^p\|^2, \qquad y(A^{l-1}) = \|A^{l-1}:X^p\|^2,$$

and observe that

$$\|A^l:X^p\| \leqq \|A^{l-1}:X^p\| \qquad \text{for } \{A^{l-1}\} \subset \{A^l\};$$

therefore by definition and (10.6)

$$c \leqq \binom{h}{l-1} = \binom{h}{l+1}, \qquad x \leqq \binom{h}{l+1}y.$$

This coincides with the desired inequality

$$d_i^2 \leqq \binom{h}{i+1} d_{i+1}^2 .$$

In the above argument leading to (10.6) one may interchange the roles of $x(A^l)$ and $y(A^{l-1})$ whence this double result:

LEMMA 10.B. With the notations (10.5) one has

$$\frac{x}{y} \leqq \frac{1}{h-l+1} \sum \frac{x(A^l)}{y(A^{l-1})}, \qquad \frac{y}{x} \leqq \frac{1}{l} \sum \frac{y(A^{l-1})}{x(A^l)} .$$

Let $\{X^p\}$ be given. We wish to describe the manifold of all elements $\{A^h\}$ which are in i-incidence with $\{X^p\}$. In this study the unitary metric is immaterial. Suppose X^p to be the unit p-ad $E^p = [e_1 \ldots e_p]$ and write down the matrix of the coordinates of the vectors $a_1, \ldots, a_h$ which span $A^h \neq 0$,

$$\left\| \begin{array}{ccc|ccc} a_{11}, & \ldots, & a_{1p} & a_{1,p+1}, & \ldots, & a_{1n} \\ \cdot & \cdot & \cdot & \cdot & \cdot & \cdot \\ a_{h1}, & \ldots, & a_{hp} & a_{h,p+1}, & \ldots, & a_{hn} \end{array} \right\| .$$

It splits into a front and a rear part of p and n - p columns respectively. The vectors $a_1, \ldots, a_h$ mod E^p are the rows of the rear part. That $\{A^h\}$ is in i-incidence with $\{E^p\}$ means that this rear part is of rank $< l = h - i$, or that any l columns out of it are linearly dependent. Hence the determinant of h columns selected from the whole matrix is zero if at least l columns belong to the rear, or

(10.7) $$A^h(i_1, \ldots, i_h) = 0$$

whenever at least l of the indices $i_1, \ldots, i_h$ belong to the range $p + 1, \ldots, n$. Vice versa if these equations are satisfied, then any l columns of the rear part are linearly dependent. Indeed, were we able to find l among them that are linearly independent, we could supplement them by h - 1 further columns from the whole scheme, so that linear independence is maintained, contrary to the equations (10.7). In the $\binom{n}{h}$-space $\mathfrak{S}^{(h)}$ of all h-ads these equations define a linear subspace $\mathfrak{E}_i^p$, and we have proved that an h-element $\{A^h\}$ is in i-incidence with $\{E^p\}$ if and only if the special h-ad A^h belongs to this subspace $\mathfrak{E}_i^p$. The non-singular linear transformations in $\mathfrak{S}$ induce certain transformations in the space $\mathfrak{S}^{(h)}$, see (I, 1.4); two linear subspaces in $\mathfrak{S}^{(h)}$ are considered equivalent only if they are carried into one another by one of these induced transformations. Our result may be stated as follows: With any $\{X^p\}$ there is associated a (uniquely determined) linear subspace $\mathfrak{X}_i^p$ of $\mathfrak{S}^{(h)}$ such that the element $\{A^h\}$ is in i-incidence with $\{X^p\}$ if and only if the special h-ad A^h belongs to $\mathfrak{X}_i^p$. This subspace $\mathfrak{X}_i^p$ is equivalent to $\mathfrak{E}_i^p$.

Let us compute the dimensionality W_i^p of $\mathfrak{E}_i^p$. Any h-ad belonging to it is characterized by the ordered components $A(i_1 i_2 \ldots i_h)$ of whose indices $i_1, \ldots, i_h$ not more than $i - 1$ are in the range $p + 1, \ldots, n$; these components are independent, while the others vanish. The number of combinations $i_1 < i_2 < \ldots < i_h$ in which exactly the last k elements belong to the range $p + 1, \ldots, n$ is $\binom{p}{h-k} \cdot \binom{n-p}{k}$. Hence

$$(10.8)\qquad \begin{aligned} \sum_{k=0}^{h} \binom{p}{h-k}\binom{n-p}{k} &= \binom{n}{h}, \\ W_i^p &= \sum_{k=0}^{h-i-1} \binom{p}{h-k}\binom{n-p}{k}. \end{aligned}$$

After these preparations we are going to <u>formulate the general defect relation</u>. We start with a given rank number h. Suppose we are given a finite set $\sum$ of h-elements $\{A\}$, $A = A_1, \ldots, A_q$, a positive weight $w(A)$ being atrached to each of them. We denote the total weight $\sum w(a)$ by w. Suppose further that the sum of the weights of those among the q given $\{A\}$ which are in i-incidence with a p-element $\{X^p\}$ never exceeds the value w_i^p whatever

$\{X^p\}$. In order to find the least w_i^p satisfying this condition, one picks out an arbitrary combination from the q elements $\{A\}$, examines whether there is a p-element $\{X^p\}$ with which all the $\{A\}$ of that combination are in i-incidence, and if so forms the sum of their weights. w_i^p is the largest of the sums thus obtained.

<u>GENERAL DEFECT RELATION (THIRD MAIN THEOREM)</u>:

$$\sum_{\{A^h\}\in\Sigma} w(A^h)\{\tilde{m}_p(A^h)-\tilde{m}_{p-1}(A^h)\}+(w_0^p\Omega_p+\dots+w_{h-1}^{p+h-1}\Omega_{p+h-1}) \tag{10.9}$$
$$= \tfrac{1}{2}(w_0^p+\dots+w_{h-1}^{p+h-1})\cdot\omega(T^2)$$

$$(p + h \leqq n).$$

Here T^2 stands for a linear combination $B_0T_p^2 + \dots + B_{h-1}T_{p+h-1}^2$ with certain constant positive coefficients $B_0, \dots, B_{h-1}$ which are determined by the configuration Σ and the attached weights.

Let us say that the set Σ is in general position if none of the linear subspaces $\mathfrak{X}_i^{p+i}$ of $\mathfrak{G}^{(h)}$ which are equivalent to $\mathfrak{C}_i^{p+i}$ contains more "points" of the system Σ than its dimensionality W_i^{p+i} indicates, and this for $i = 0, 1, \dots, h-1$. Set $w(A) = 1$. Under the hypothesis of general position we may then choose the dimensionalities W_i^{p+i} given by (10.8) as our w_i^{p+i}, and the defect relation assumes the form

$$\sum\{\tilde{m}_p(A^h) - \tilde{m}_{p-1}(A^h)\} + (W_0^p\Omega_p+\dots+W_{h-1}^{p+h-1}\Omega_{p+h-1}) \tag{10.10}$$
$$= \tfrac{1}{2}(W_0^p+\dots+W_{h-1}^{p+h-1})\cdot\omega(T^2).$$

§11. Proof of the general defect relation

In the inequality (6.6) and the expression (6.4) of $F(A^h)$ we replace h by $l - 1$ and assume b perpendicular to $\{A^{l-1}\}$, so that A^{l-1} and $A^l = [b,A^{l-1}]$ are special $(l-1)$-

and l-ads normalized by $|A^{l-1}| = |A^l| = 1$. Moreover we replace $T_p(A^{l-1})$ by the larger $T_p + \tilde{m}^0_p(A^{l-1}) = T_p + (o)$. We thus have the inequality

$$\frac{1}{\pi}\int \phi \cdot \frac{|[A^{l-1}X^{p+1}]|^2 \cdot |[A^l X^{p-1}]|^2}{|[A^{l-1}X^p]|^2 \cdot |[A^l X^p]|^2} \cdot J^*\left(\frac{|[A^l X^p]|^2}{|[A^{l-1}X^p]|^2}\right) \leqq$$

$$\leqq e_{n-p-l+1}(T_p + (o)).$$

According to condition (III) the value of J^* is lowered if its argument is replaced by the smaller $\frac{|[A^l X^p]|^2}{|X^p|^2}$.

Let now $A = A^h = [a_1 \ldots a_h]$ be a fixed special p-ad spanned by a normal basis, and then sum over the $\{A^{l-1}\}$, $\{A^l\}$, $\{A^{l-1}\} \subset \{A^l\}$, spanned by vectors of that basis, in the same manner that we formed the sum (10.3); $l = h-i$. On the right side we have to add the factor $l \cdot \binom{h}{l}$. Since $J^* \leqq 1$ the sum is further decreased by replacing each individual factor $J^*(\|A^l : X^p\|^2)$ by

$$f^p_i(A) = \prod J^*(\|A^l : X^p\|^2),$$

the product extending over the $\binom{h}{l}$ elements (10.1). At this stage we are dealing with the expression

$$\frac{|X^{p-1}|^2 \cdot |X^{p+1}|^2}{|X^p|^4} f^p_i(A) \cdot \sum \frac{\|A^{l-1} : X^{p+1}\|^2 \cdot \|A^l : X^{p-1}\|^2}{\|A^{l-1} : X^p\|^2 \cdot \|A^l : X^p\|^2}$$

under the sign of integration. To the sum the second part of Lemma 10.B is applied with

$$y(A^{l-1}) = \frac{\|A^{l-1} : X^{p+1}\|^2}{\|A^{l-1} : X^p\|^2} \quad \text{and} \quad x(A^l) = \frac{\|A^l : X^p\|^2}{\|A^l : X^{p-1}\|^2}.$$

Setting

$$\sum_{A^l} \frac{\|A^l : X^p\|^2}{\|A^l : X^{p-1}\|^2} = \binom{h}{l} \cdot \Psi^p_i(A) \qquad (i = h-l)$$

we obtain

$$\frac{1}{2\pi}\int \phi S^p \cdot F_i^p(A) \leqq \binom{h}{l} e_{n-p-l+1}(T_p+(o))$$

where

$$F_i^p(A) = f_i^p(A)\cdot\frac{\Psi_{i+1}^{p+1}(A)}{\Psi_i^p(A)}.$$

Finally, replace p by p + i and for the sake of brevity write f_i, Ψ_i instead of f_i^{p+i}, Ψ_i^{p+i}:

$$\frac{1}{2\pi}\int \phi S^{p+i} f_i(A)\frac{\Psi_{i+1}(A)}{\Psi_i(A)} \leqq \binom{h}{i} e_{n-p-h+1}(T_{p+i}+(o)).$$

Suppose now positive weights w(A) attached to the elements $\{A\} = \{A^h\}$ of a given finite set $\sum$ and let $w_i^{p+i} = w_i$ have the significance described above. Each A of $\sum$ is referred to a normal basis. Notice that $f_i \leqq 1$, $\Psi_{i+1} \leqq 1$ and

$$\binom{h}{i}\Psi_i(A) \geqq \sum \| A^l : X^{p+i} \|^2 \quad \text{or} \quad \Psi_i(A) \geqq d_i^2(A, X^{p+i}),$$

d_i being the measure of i-incidence as introduced in the previous section. Repeating the proof of Lemma 7. A. we obtain a relation

$$\exp\left\{\frac{1}{w_i}\sum_A w(A)\log\left(f_i(A)\frac{\Psi_{i+1}(A)}{\Psi_i(A)}\right)\right\}$$

$$\leqq C\left\{1 + \frac{1}{w_i}\sum_A w(A) f_i(A)\frac{\Psi_{i+1}(A)}{\Psi_i(A)}\right\}$$

where C is a constant which depends on nothing but the configuration $\sum$ with its distribution of weights w(A). The cell $\mathfrak{Z}(t_1,\ldots,t_q)$ is here to be defined by the inequalities

$$d_i(A_{t_1}, X^{p+i}) \leqq d_i(A_{t_2}, X^{p+i}) \leqq \ldots \leqq d_i(A_{t_q}, X^{p+i}).$$

We arrive at the relation

$$\frac{1}{2\pi}\int \phi S^{p+i} \cdot \exp\{\frac{1}{w_i}\sum w(A)(\log f_i(A)+\log\Psi_{i+1}(A)-\log\Psi_i(A))\}$$

$$\leqq C\{1 + \frac{w}{w_i}\binom{h}{i}e_{n-p-h+1}\}(T_{p+i}+(o)).$$

Set

$$2M^{(i)}(r;A) = -\frac{1}{2\pi}\int_{\Gamma_r} \log\Psi_i(A)\cdot d\vartheta,$$

$$2\varepsilon^{(i)}(r;A) = -\frac{1}{2\pi}\int_{\Gamma_r} \log f_i(A)\cdot d\vartheta.$$

Then the equation

$$\int_0^R (R-r)\cdot e^{2\Theta^{(i)}(r)}dr \leqq C\{1 + \frac{w}{w_i}\binom{h}{i}e_{n-p-h+1}\}(T_{p+i}+(o))$$

results where

$$\Theta^{(i)}(r) = \Pi^{(i)}(r) - \frac{1}{w_i}\sum w(A)\varepsilon^{(i)}(r;A),$$

$$\Pi^{(i)}(r) = \Omega_{p+i}(r) + \frac{1}{w_i}\sum w(A)\{M^i(r;A) - M^{(i+1)}(r;A)\}.$$

We now set $J(s) = s^{-\lambda}$ $(\lambda < 1)$ which gives

$$\varepsilon^{(i)}(r;A) = (1-\lambda)\sum \tilde{m}_{p+i}(r;A^l),$$

the sum extending over the elements (10.1) derived from the normal basis $a_1, \ldots, a_h$ of A. Denote by $\tilde{m}_0^{(i)}[A]$ the corresponding sum $\sum m^o_{p+i}(A^l)$. Then

$$\varepsilon^{(i)}(r;A) \leqq (1-\lambda)\{\binom{h}{l}T_{p+i}(r) + \tilde{m}_0^{(i)}[A]\}$$

$$\leqq (1-\lambda)\{\binom{h}{i}T_{p+i} + \tilde{m}_0^{(i)}[A]\}.$$

We therefore choose λ in terms of G according to the equation

$$\frac{2}{w_i}\{w\binom{h}{i}T_{p+i} + \sum_A w(A)\tilde{m}_0^{(i)}[A]\} = (1-\lambda)^{-1}$$

and noticing that $\tilde{m}_0^{(i)}[A] = (o)$ arrive at the formula

$$(11.1)\quad \Pi^{(i)} = \Omega_{p+i} + \frac{1}{w_i}\sum_A w(A)\{M^{(i)}(A) - M^{(i+1)}(A)\} =$$

$$= \frac{1}{2}\omega(B_i T^2_{p+i})$$

$$(i = 0,\ldots,h-1)$$

where B_i is any number greater than

$$2eC(\frac{w}{w_i})^2\cdot(\frac{h}{i})^2.$$

Observe that

$$(11.2)\qquad \Psi_0(A) = \frac{\|A:X^p\|^2}{\|A:X^{p-1}\|^2},$$

and hence $M^{(0)}(A) = \tilde{m}_p(A) - \tilde{m}_{p-1}(A)$. The relation (11.1) is of recursion type with respect to i. We multiply by $\frac{w_i}{w_0+\ldots+w_{h-1}}$ and then, summing with respect to i from 0 to $h-1$, by means of Lemma 10. A. arrive at the preannounced result

$$(11.3)\quad \begin{aligned}&\sum_A w(A)\{\tilde{m}_p(A) - \tilde{m}_{p-1}(A)\} + \sum_{i=0}^{h-1} w_i\Omega_{p+i}\\ &= \frac{1}{2}(w_0+\ldots+w_{h-1})\cdot\omega\left(\frac{w_0B_0T_p^2+\ldots+w_{h-1}B_{h-1}T^2_{p+h-1}}{w_0+\ldots+w_{h-1}}\right).\end{aligned}$$

The boldness of this procedure lies in the introduction of the artificial quantity $\Psi_i(A)$ which in general is not unitarily invariant, but for $i = 0$ reduces to the invariant (11.2). We should feel happier if instead of $\Psi_i(A)$ the invariant quantities

$$\frac{\sum\|A^l:X^p\|^2}{\sum\|A^l:X^{p-1}\|^2}$$

appeared; but it did not come out that way.

The relation dual to (10.9) reads:

$$\sum w(A)(\hat{m}_{n-p}(A) - \hat{m}_{n-p+1}(A)) + (w_0^p\Omega_{n-p}+\dots+w_{h-1}^{p+h-1}\Omega_{n-p-h+1})$$

(11.4)
$$= \tfrac{1}{2}(w_0^p+\dots+w_{h-1}^{p+h-1})\cdot\omega(T^2)$$

where this time T^2 stands for a linear combination of $T^2_{n-p}, \dots, T^2_{n-p-h+1}$.

§12. Computation of coefficients. The fundamental relation for the actual defects

We may substitute $V_p + (T_{p+1} - 2T_p + T_{p-1})$ for Ω_p in (10.9) if $\frac{1}{2}\omega(T^2)$ in the right member is simultaneously replaced by $\eta + \frac{1}{2}\omega(T^2) + (o)$. The relation thus obtained remains true a fortiori after dropping the V's. For the moment denoting $T_p - T_{p-1}$ by ΔT_p we transform the sum

$$\sum_{i=0}^{h-1} w_i^{p+i}(T_{p+i+1} - 2T_{p+i} + T_{p+i-1}) = \sum_{i=0}^{h-1} w_i^{p+i}(\Delta T_{p+i+1} - \Delta T_{p+i})$$

by partial summation into

(12.1)
$$\sum_{i=0}^{h}(w_{i-1}^{p+i-1} - w_i^{p+i})\Delta T_{p+i}$$

where $w_{-1}^{p-1} = w_h^{p+h} = 0$. Let us compute the coefficients of the latter sum for the case of general position with

$$w(a) = 1, \qquad w_i^{p+i} = W_i^{p+i}, \qquad (i = 0,\dots,h-1).$$

The formula (10.8) gives $W_h^{p+h} = 0$ in agreement with the convention $w_h^{p+h} = 0$, but $W_{-1}^{p-1} = \binom{n}{h}$ instead of 0. Hence (12.1) takes on the form

$$\sum_{i=0}^{h}(W_{i-1}^{p+i-1} - W_i^{p+i})\Delta T_{p+i} - \binom{n}{h}\Delta T_p.$$

By definition

$$W_i^{p+i} - W_{i-1}^{p+i-1} =$$

$$\sum_{k=0}^{h-i}\left\{\binom{p+i}{h-k}\binom{n-p-i}{k} - \binom{p+i-1}{h-k}\binom{n-p-i+1}{k}\right\} - \binom{p+i}{i}\binom{n-p-i}{h-i}.$$

Because of the recursive formula

$$(12.2) \qquad \binom{n+1}{i} = \binom{n}{i} + \binom{n}{i-1}$$

for the binomial coefficients, the individual term of the last sum may be transformed into

$$\left\{\binom{p+i}{h-k} - \binom{p+i-1}{h-k}\right\}\binom{n-p-i}{k} - \binom{p+i-1}{h-k}\left\{\binom{n-p-i+1}{k} - \binom{n-p-i}{k}\right\}$$

$$= \binom{p+i-1}{h-k-1}\binom{n-p-i}{k} - \binom{p+i-1}{h-k}\binom{n-p-i}{k-1}.$$

This is of the form $f(k+1) - f(k)$ and, as $f(0) = 0$, the sum over k from 0 to $h - i$ gives $f(h-i+1)$, i. e.

$$\binom{p+i-1}{i-1}\binom{n-p-i}{h-i},$$

and once more applying (12.2) we get

$$W^{p+i-1}_{i-1} - W^{p+i}_{i} = \binom{p+i-1}{i}\binom{n-p-i}{h-i}.$$

Thus

$$(12.3) \qquad \begin{aligned} &\sum_{A^h} \Delta\tilde{m}_p(A^h) + \left\{\sum_{i=0}^{h}\binom{p+i-1}{i}\binom{n-p-i}{h-i}\cdot \Delta T_{p+i} - \binom{n}{h}\Delta T_p\right\} \\ &= (W^p_0 + \ldots + W^{p+h-1}_{h-1})\{\eta + \tfrac{1}{2}\omega(T^2) + (0)\}, \end{aligned}$$

and this relation contains the complete result of our investigation as far as systems of h-elements $\{A^h\}$ in general position are concerned. As in §9 the exponential curves can be used as an illustration showing that the result cannot be improved by changing the coefficients of the ΔT.

If a defect relation is wanted involving $\tilde{m}_p(A)$ rather than the difference $\Delta\tilde{m}_p = \tilde{m}_p - \tilde{m}_{p-1}$, one replaces p in (10.9) by p' and then sums over $p' = 1,\ldots,p$ by means of Lemma 10. A. Use the summation letter $p' + i = s$; then a formula is obtained,

$$\sum_A w(A)\tilde{m}_p(A) + \sum_{i,s} w_i^s \Omega_s = \frac{1}{2}\sum_{i,s} w_i^s \cdot \omega(T^2),$$

in which T^2 is a linear combination of $T_1^2, \ldots, T_{p+h-1}^2$ and the sums (i,s) are double sums over the range

$$(12.4) \qquad 0 < s - i \leqq p, \qquad 0 \leqq i < h.$$

We are going to carry out this summation for general position, with $w(A) = 1$, $w_i^s = W_i^s$. Hence we assume that for the i and s satisfying (12.4) no $\mathfrak{X}_i^s$ contains more than W_i^s of the elements $A = A^h$ of the set $\sum$, or that there is no $\{X^s\}$ with which more of them are in i-incidence. What we have to do is to replace p by p' in (12.3) and then to sum over $p' = 1, \ldots, p$. But before doing so let it be observed that the sum $\sum_{i=0}^h$ may formally be written as a sum extending over i from $-\infty$ to $+\infty$, provided we define the binomial coefficients for all i by the identity in x,

$$(1+x)^n = \sum_{i=-\infty}^{+\infty} \binom{n}{i} x^i.$$

Summation by parts transforms the sum over i in (12.3) into

$$\sum_{i=-\infty}^{+\infty} \left\{ \binom{p+i-1}{i}\binom{n-p-i}{h-i} - \binom{p+i}{i+1}\binom{n-p-i-1}{h-i-1} \right\} T_{p+i},$$

and the expression resulting from summation over $p' = 1, \ldots, p$ is

$$(12.5) \sum_A \tilde{m}_p(A) + \sum_{i,s} \left\{ \binom{s-1}{i}\binom{n-s}{h-i} - \binom{s}{i+1}\binom{n-s-1}{h-i-1} \right\} T_s - \binom{n}{h} T_p.$$

Again the sum (i,s) is a double sum over i and s with the range

$$i < s \leqq i + p \quad \text{or} \quad s - p \leqq i < s.$$

First carry out summation with respect to i; one realizes at once that the upper limit $s - 1$ may formally be replaced by $+\infty$. Thus we have to calculate

$$(12.6)\qquad \sum_{i=s-p}^{\infty}\{\binom{s-1}{i}\binom{n-s}{h-i} - \binom{s}{i+1}\binom{n-s-1}{h-i-1}\}.$$

In this case it is a little easier to utilize the recursion relations for the binomial coefficients by means of their generating function. $\binom{s-1}{i}\binom{n-s}{h-i}$ is the coefficient of $t^i x^h$ in the polynomial $(1+tx)^{s-1}(1+x)^{n-s}$. Hence

$$\sum_{i=s-p}^{\infty}\sum_{h=0}^{n-1}\{\binom{s-1}{i}\binom{n-s}{h-i} - \binom{s}{i+1}\binom{n-s-1}{h-i-1}\}x^h$$

is the coefficient of t^0 in the Laurent series of the following function of t:

$$\sum_{i=s-p}^{\infty} t^{-i}\{(1+tx)^{s-1}(1+x)^{n-s} - t^{-1}(1+tx)^{s}(1+x)^{n-s-1}\}$$

$$= \frac{t^{-(s-p)}}{1-t^{-1}}(1+tx)^{s-1}(1+x)^{n-s-1}\{(1+x) - t^{-1}(1+tx)\}$$

$$= t^{-(s-p)}(1+tx)^{s-1}(1+x)^{n-s-1}.$$

Thus the sum (12.6) equals

$$\binom{s-1}{s-p}\binom{n-s-1}{h-s+p},$$

and the expression (12.5) changes into

$$(12.7)\qquad \sum_A \tilde{m}_p(A) + \sum_{s=p}^{p+h}\binom{s-1}{s-p}\binom{n-s-1}{h-s+p}T_s - \binom{n}{h}T_p.$$

It is this combination which we know is essentially negative, provided the order T satisfies the hypothesis $\mathfrak{H}$ and the h-elements $\{A\}$ are in general position.

The result is of particular interest for $h = n - p$, because $\tilde{m}_p(A^{n-p})$ is the compensating term $m_p(A^p)[A^p = {}^*A^{n-p}]$ which appears in the First Main Theorem, and hence is a real defect, not merely a defect of a defect. What we find is the following

THEOREM. Under the assumption of general position and under the hypothesis $\mathfrak{H}$ the inequality

$$(12.8) \quad \| \quad \sum_{A^p} m_p(A^p) < \{\binom{n}{p} + \epsilon\}T_p$$

holds for almost all G, however small the preassigned positive number ϵ; in particular (p = 1)

$$(12.9) \quad \| \quad \sum_{\alpha} m_1(\alpha) < (n + \epsilon)T_1 .$$

General position for the set $\sum$ of contravariant p-elements $\{A^p\}$ certainly prevails if they satisfy no accidental linear relation whatsoever, i. e. if no linear subspace of the space $\mathfrak{G}^{*(p)}$ of all contravariant p-ads contains more elements of $\sum$ than its dimensionality indicates. For p=1 our condition of general position is equivalent to this sweeping requirement, whereas for higher p it demands much less. We add the following observation for the case p = 1. Provided the total number q of planes (α) exceeds n, the set of conditions that there is no i-element of $\mathfrak{G}$ which goes through more than n - i planes (α) of $\sum$, stipulated for i = 1, ..., n - 1, can be reduced to the one condition i = 1, that there is no point through which more than n - 1 of the planes (α) pass. In the opposite case $q \leq n$ the relation (12.9) is trivial, whatever the position of the q planes (α).

The results (12.8) and (12.9) are of the same beauty and simplicity as Picard's theorem and Nevanlinna's defect relation for meromorphic functions. One is perhaps justified in looking upon the whole development of this chapter and its other results as mere means to this end. One could wish for a more direct approach. However, one must not forget the condition of general position; if that condition actually reflects the natural limitation under which the law (12.8) holds, it would go far to explain why this law is not as easily accessible as its simplicity seems to promise.

In particular, we have the following generalization of Picard's theorem, which in the z-plane is equivalent to a theorem of E. Borel's known for a long time:

Let there be given any $n + 1$ *planes in n-space in general position (i. e. no* n *of which go through a common point) and an analytic curve the order of which fulfills the hypothesis* $\mathfrak{H}$. *Then the curve will cut at least one of the planes.*